EXTRACT GOLD

Chemical Process From Gold Plated Electronics
Pins

(Tricks and Techniques About the Process of
Recovering Gold)

Norman Swasey

I0049162

Published By **Norman Swasey**

Norman Swasey

All Rights Reserved

Extract Gold: Chemical Process From Gold Plated Electronics Pins (Tricks and Techniques About the Process of Recovering Gold)

ISBN 978-1-77485-437-2

All rights reserved. No part of this guide may be reproduced in any form without permission in writing from the publisher except in the case of brief quotations embodied in critical articles or reviews.

Legal & Disclaimer

The information contained in this book is not designed to replace or take the place of any form of medicine or professional medical advice. The information in this book has been provided for educational and entertainment purposes only.

The information contained in this book has been compiled from sources deemed reliable, and it is accurate to the best of the Author's knowledge; however, the Author cannot guarantee its accuracy and validity and cannot be held liable for any errors or omissions. Changes are periodically made to this book. You must consult your doctor or get professional medical advice before using any of the suggested remedies, techniques, or information in this book.

Upon using the information contained in this book, you agree to hold harmless the Author from and against any damages, costs, and expenses, including any legal fees potentially resulting from the application of any of the information provided by this guide. This disclaimer applies to any damages or injury caused by the use and application, whether directly or indirectly, of any advice or information presented, whether for breach of contract, tort, negligence, personal injury, criminal intent, or under any other cause of action.

You agree to accept all risks of using the information presented inside this book. You need to consult a professional medical practitioner in order to ensure you are both able and healthy enough to participate in this program.

TABLE OF CONTENTS

INTRODUCTION 1

CHAPTER 1: EXPLAINS WHY ONE SHOULD INVEST IN
SILVER AND GOLD ? 3

CHAPTER 2: GETTING STARTED 23

CHAPTER 3: NUTS AND BOLTS 33

CHAPTER 4: SYSTEMS AND PROCESSES SYSTEMS 46

CHAPTER 5: THE CLIENT 63

CHAPTER 6: THE WHOLE PICTURE 71

CHAPTER 7: WHAT TO CONSIDER ABOUT 80

CHAPTER 8: QUESTIONS AND RESPONSES 89

CHAPTER 9: THE INACTIVE LOOK FOR SILVER AND GOLD 95

CHAPTER 10: WHAT IS THE BEST TIME TO PURCHASE
GOLD, AND WHAT IS THE BEST TIME TO DO SO? I PREFER
TO PURCHASE SILVER? 110

CHAPTER 11: THE URBAN VALLEY PROSPECTORS 122

CHAPTER 12: THE THREE 9V BATTERY COLLOIDAL SSILVER
MAKER 163

CHAPTER 13: MAKE SWALLOW SILVER BY COMBINING THE
RECENT-DISTILLATED WATER (2 POT APPROACH) 168

CHAPTER 14: 6 MATTERS TO KEEP IN MIND BEFORE DYING
YOUR DIY COLOIDAL SILVER GENERATOR 172

CHAPTER 15: WHEN IS THE IDEAL TIME TO PURCHASE SILVER AND GOLD FROM PRIVATE INDIVIDUALS? 179

CONCLUSION 181

Introduction

I would like to express my gratitude for your benefit in this particular distribution. Chances are that you're looking for reliable information on the techniques used to recover valuable gold and various metals. So have I. In the past ten years, I have seen a lot of progress. From normal E-waste to devices I have witnessed being thrown away and destroyed. I realized in all the ways that everything was gold. However, I was told that it's too difficult or dangerous. Well after many years of exploration and numerous endeavors and just seeing it completed was an awakening. It opened my eyes and led me to realize that there's a gold mine that is waiting to be mined!

Just before us. Therefore, I've decided to write a book in order to provide what I've viewed as the most simple and cheapest method of reclaiming the gold in esquander. I had to make it clear in a step-by-step quick and simple way. In this article, I will explain five different ways of removing gold. The easiest method to start with first. Information that is not garbage or wasteful substance. The reader is not wasting cash and time. This book will provide the reader everything needed to start.

This is the only one you'll need to have at any time. It's beginning and ending as well as other subtleties regarding data. Start your collaborations. It is advisable to begin with someone who is familiar with this, or requires some money to learn by yourself. Additionally, make the best of the information which has been made clear in this distribution. How can we start rolling back gold!

Chapter 1: Explains Why One Should Invest In Silver And Gold ?

The question is simple to answer. Silver and gold are both the two most valuable currencies. Physical possessions have always and are still worth it. The value of silver and gold diminished. At the time of Rome one could purchase an toga for a pound of gold (31.1 grams) and today, an elegant suit for men. This is a proof that gold has held its value throughout many thousands of years.

In particular, in a time when central banks, with their phrases of whatever they want or with no limit, make money out of nothing and it becomes even more essential to own the precious metals physical.

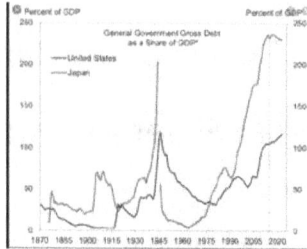

Percent of GDP

General Government Gross Debt
as a Share of GDP*

— United States
— Japan

The reason behind the growing appeal of gold is evident: "The overriding discussion is the negative interest rates" According to Benjamin Summa, spokesman of Pro Aurum, a Munich gold trading company Pro Aurum. "Many bank and saving banks charge interest to savings deposits for a while currently, and in the meantime , there are institutions that pay interest to their customers from the very first euro up to. If inflation is taken into consideration the savings account holders lose money in the absence of penalty. However, gold has always been regarded as an investment that is relatively safe. "Before the end of the year, we were selling out fast and there were massive supply issues," says Summa. Gold has always held the highest value "Measured by the general

4

population, Germany is the country that has the greatest affinity to the precious metal," says BayernLB chief trader Eubel. As per the study, Germans purchase around 100 tons of shiny metal each year. Additionally to that, the Bundesbank is the second-largest gold reserve which total over 3300 tonnes.

Silver has more potential for appreciation than gold

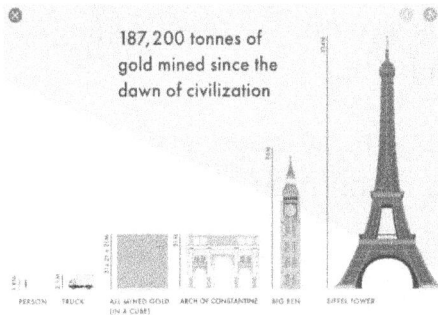

However, it is not wise to overlook silver during"the "gold rush". Degussa director of economics Thorsten Polleit is convinced that when the precious metal isn't at the forefront, it stands an excellent chance of increasing in value. "Gold is the best method of human payment," he explains in an interview with FOCUS Online. "It has served as a payment for over 4000 years

Its function as a currency and I am sure that this will be so in the coming years. But, silver may gain more value than gold in certain instances the economist said. The price of gold has seen a sharp increase, however, the price of silver hasn't yet increased. "Therefore according to my opinion the upside potential of silver is greater in the coming one and two-and-a-half years.

2. What are my options to purchase silver and gold other than through the dealer?

Starting in the year 2020 Germans are no longer able buy gold for anonymously for 10,000 euros. Instead, they will be able to buy it for less than two thousand euros. The move by the Federal Government has made big waves that have gone beyond the typical gold and silver trade. Many experts believe it's only the beginning of the end of the road before anonymous cash transactions when purchasing silver and gold will be banned completely.

Coronavirus: Supply bottlenecks for gold, the precious metal as well as similar news stories are regularly appearing on the news. But in reality, it's becoming increasingly difficult to find physical silver and gold. If you visit a dealer in precious metals or dealer, they are either closed or only offers online delivery with extremely long time to deliver. When you consider the fees associated with the gold price in the sport (gold market price, which is traded through the Stock Exchange) is significant. For silver, a surcharge of up to 100% is payable.

In the end, private investors are unable to purchase at market prices any longer. A price comparison that is current is recommended.

Alongside purchasing precious metals from dealers, it is recommended to purchase from firms that handle precious metals. The benefit is that you can buy any metal without duty. This is especially beneficial for the metals platinum, palladium, and silver as you do not have to pay tax on value added. These are some of the most interesting information about precious metals administrators.

Apart from purchasing precious metals directly from the traders, it's recommended to purchase from firms that handle precious metals. This is because you can purchase all the metals for free. This is especially beneficial for precious metals such as platinum, palladium, and silver as you do not have to pay tax on value added. You can purchase so known as vault gold

It is possible to acquire. The price of purchase and selling are sometimes as high as 0.10 percent. Vault gold is the least expensive and most secure form of investment in gold, since it's among other things. ,...

The fund does not simply replicate the price of gold as an ETF, it is actually assigned physical gold (i.e. real gold, as opposed to gold derivatives, aka "paper gold")

Meets the requirements of the professional market for gold and is thus internationally recognized and is traded throughout the world.

In contrast to securities, it's not a bond and is 100% ownership by the investor

is in a vault that is insured by professional vault operators.

Here are some fascinating precious metal administrators.

BullionVault: There are a variety of Bonded warehouses that are available for gold in Zurich, London, New York, Toronto, Singapore for syllables for Zurich, London, Toronto and Singapore and platinum: London

Join for no cost here:

Goldmoney This company Goldmoney also provides the possibility of storing precious metals.

Alongside the physical purchase of silver and gold via dealers and managers of precious metals There is also the possibility of investing in shares. There is a distinction between purchase of mining shares or the shares of precious metal services providers.

The shares of service providers have a lower fluctuation margin as compared to the pure mine operators or Mini Explorer shares, which means have a lower risk.

Here are some intriguing services for precious metals:

Goldmoeny Inc.

The company is a metals payment network, and provides investment options for precious metals. The objective is to make savings backed by precious metals available to all. Goldmoney Holding, based on the company's own patented technology it is an online account that lets customers make investments, earn or use platinum, gold, silver as well as crypto currencies like palladium. kept in vaults that are insured across more than six countries. The value of each bar is completely allotted and can be physically exchanged. Goldmoney Wealth Limited is regulated by the Jersey Financial Services Commission (JFSC) as a money service business. Goldmoney Network is a reporting part that is part of the Financial Transactions and Reports Analysis Centre of Canada (FINTRAC) and is registered with the Financial Crimes Enforcement Network (FinCEN) in the United States.

Wheaton precious metals

Silver Wheaton is a Canadian trading company for precious metals, with its headquarters in Vancouver. It is involved in what is known as "silverstreaming" which is included on the index of shares S&P/TSX. Silverstreaming refers to the process whereby some or all of the production of gold or silver of mining companies are bought to be sold for a set price. Mining companies receive an assurance of sales. The Silver-Wheaton-Corporation claims to be the world's largest silverstreaming company and generates 60% of its revenues from silver and 40% from gold sales. The company was founded in 2004. was established as a spin-off from Canadian mining company Goldcorp. Up to the 7th of December, 2006 Goldcorp held 48 percent of Silver Wheaton. On February 14, 2008, Goldcorp has sold the remaining shares.

Franco-Nevada

Franco-Nevada Corporation is a Canadian mining company based in Toronto, Ontario. It has investments in mines as well as the production of metals, including gold, and also natural gas and oil. Franco-Nevada is not operating any mines for itself. The revenue is generated by licensing agreements as well as shares in the extracted raw material (streams). The objective is to ensure that 80percent of the revenue are derived out of precious metals. The shares of the Company trade on both the Toronto Stock Exchange and New York Stock Exchange and are part of the TSX60 share index.

Dealing with precious metal management companies and purchasing shares of a service provider that deals in precious metals is the best option when you don't wish to own the silver and gold at your home. However, if you wish to keep silver and gold at home, buying of precious metals from the commercial market has in a rut in the wake of when there was the Corona crisis.

Private trade, or the private acquisition of silver and gold isn't well-known but it is a great opportunities.

3. The solution to the purchase of private from private.

The buying and selling of silver and gold in any form require some experience. With the advent of digitalization and the application of various testing methods whether it's the traditional older acid residue, or testing using electronic instruments for measuring It has become more simple to buy and sell silver and gold privately. Thus, one can guard yourself against fraud. There is a whole chapter on this.

In general, these points should be taken into consideration that gold coins and bars have a value that is generally more than the current price of gold. The so-called market value gets decreased if the gold bars or coins are damaged. The coins or bars may be scratched or even tiny edges could be damaged. The dealer is then faced with problems selling the coins or bars to new customers. They're no longer tradeable. For these pieces, the seller can provide the purchase price along with discounts. What is the situation?

If a currency is scratched to the point that it is unable to be traded, it is important to distinguish between coins made from pure gold and coins that are made from an alloy. For alloys, we have the melting pricebecause it has to be re-used and separated like the old jewelery. If it's made from fine gold, the higher quality crystals are priced because it doesn't have to be divided, but rather melting again.

Silver, which is tarnishing, on bars and coins is a frequent issue. Why? Our air contains hydrogen sulphide. If the silver isn't protected the silver sulphide will form. This is exacerbated by touching the pieces frequently. Some dealers only buy very heavily tarnished pieces at a discounted price. However the companies usually offer the tarnished silver coins to resell at a cheaper price than brand new coins.

Dealers who aren't scrupulous, like those that advertise on TV and radio request that you send the gold they have to. They typically pay between 10 to 20% of the gold value. Most pawnshops will pay between 30%-35 percent.

Traders who are legitimate offer between 40 to 65 percent. 40 percent to 65% could seem like a lot but keep in mind that any gold you purchase must be broken down then refined, tested, and then rolled into bars before it is able to be sold. Refiners who perform this process will offer between 85 and 92 percent of the prices of gold.

Let's examine the typical sales. A typical buyer will bring in three or four items, which weigh around 2 ounces each. The majority of the items that people bring me are 14K gold that is pure gold at 58% and the remainder are alloys that make gold more attractive and increase its hardness.

Two ounces of 58% gold yields 1.16 grams that are pure gold. Consider, for instance, that you pay this person 50 percent, which is common for all authentic gold dealer. For $1700 per ounce (which is about $850) and you then send your gold off to an refiner, who is willing to pay 95 percent of the price at the time of purchase (that is the equivalent of $1615 dollars). I'll let you perform some calculations. But it is a profitable profit in just one transaction. Numerous experts believe that gold prices will rise to $2000 for an ounce. I'm not sure if this will be the case, but just imagine the potential profits this business could be if it succeeds.

Do this three or four times per week and you'll be dealing real cash! It's legal.

4. What silver and gold do I need to buy?

Here's a list of silver and gold items that I am regularly offering.

Antiquated jewelry

bent or broken objects that are bent or broken

Gold chains

Class Rings

Bracelets

Pins

Brooch

Engagement ring or wedding ring from a former partner

Medals

watches

Twisted bracelets and chains

Wedding rings

Items with stones missing

Individual earrings

these big earrings from the 80s.

The promise ring, or jewelry from your ex-boyfriend that you'd like to forget.

If you're a female that is reading this article, I'm sure you've got one or two of these items in your jewelry or closet. If you're a male take this item to your girlfriend or wife and they'll reveal how common it is.

5. What should be taken into consideration?

This guide is aimed at the private silver and gold dealer. This guide is for those who purchase silver and gold in small amounts and then sell them in the future.

Today, I purchase from a private party like his silver and gold jewelry , and also his collection of coins and keep it for at 370 days until I can sell it back, which is tax-free since I've stored the silver and gold for longer than the 365 days. Thus, no commercial activities can be considered. It is essential to.

So, also take note of the time of the hold and the regularity. Always give them an individual collector of silver and gold jewelry, as well as silver and gold coins.

But, the definition of this as an official commercial silver and gold trader is extremely ambiguous and difficult to discern. So, be cautious and make use of the options. In the past, there was a lot of stamp collectors but only a handful did this commercially. They also made use of this view for themselves. They can also do this as the collectors of silver and gold.

A business is said as commercial when it is conducted on a regular basis, and in isolation, and with the aim to generate revenue. A license for trade can only be obtained for the activities that are covered by regulations under the Trade Licensing Act, are not banned by law, or are not morally wrong. For instance, theft from commercial premises as well as receiving stolen goods aren't covered by legislation like the Trade Regulation Act.

Only the sale and purchase of physical gold , without individual investment advice is authorized under trading trade licenses. The sale of gold certificates is subject to Securities Supervision Act in general.

6. How do you locate gold and silver sellers?

In the current crisis, many entrepreneurs and people suffer from financial issues. They are trying to find money that is liquid somehow. The government is not able to immediately take over all over the world. So, they are often forced the option of selling their sterling silverware. There are two methods to make good offers. The first is active search as well as the passive search.

a.) A.) Active lookup for private silver and gold sellers

Chapter 2: Getting Started

The first step to start an Gold Guide Business

The process of starting a business is enjoyable experience, however, it can be a bit challenging for people who had never run previously owned a business. The best thing about this type of business is its affordable initial costs and the minimal overheads to run.

The best, and at times negative, aspect of running an entrepreneur-run business is that you can set all decisions and make the shots. I think the rewards are very high in comparison to the risk. To reduce some of the risks, I believe that every small-sized business should have an initial business plan.

A well-crafted business plan can define the new business's the goals that you want to achieve, as well as the potential services that can be offered. It's a guideline which will keep you on the right path and help your business expand. Additionally, you'll require additional essential components in order to run an enterprise of a smaller size. These could include forms and a balance sheet of the income and expenditures and a cash flow statement as well as a marketing strategy, equipment list, bank account, as well as the necessary permits and licenses.

It is important to check your business plan frequently to ensure you're following what you've defined. Through reviewing your business plan, you'll be in a position to determine those areas that aren't reaching your goals and what changes and improvements need to be implemented. Make use of this data to make informed decisions as you move forward in your company.

In addition to the suggestions I provide within this guide, I suggest that you connect with other experts and make use of them as a resource to help develop and strengthen your business plan, especially when you've never had the opportunity to start an enterprise of a smaller size prior to. Numerous local and state government agencies offer great assistance to entrepreneurs who are new or existing owners. They can assist you in tackling business-related issues or at the very least point you on the right path. An excellent place to start with this is the local government's small business administration. Their goal is to aid you in this procedure, so make use of their assistance.

Do You Think This Business is Right for You?

It is vital to know that most small-sized firms fail. However, your company is not required to fall into this list. Making a few plans and work in advance is important prior to investing any amount of money. This is a cost-effective and low-cost venture that is able to be completed extremely slowly and according to your timeline. Always be aware of what's working and adjust what isn't working.

It is important to get your business right from the start. It is time to begin the process of planning your business today. Once you have your plan in place, you'll be able to start making decisions based on it. The first step is to must decide on the kind of gold guide you would like to become. Do you simply show people how to pan, or will you teach them all there is to know about mining and gold? Note down the things you wish to accomplish, what you can do as well as what you don't wish to do, and most importantly, what you can't do. Every person is unique, so the variables you list will vary between guides. When you've read this book, you'll be able to answer these questions and be more specific about these questions.

The running of a guide business demands dedication and determination. Here is a list of personal traits and aspects you need to consider and must consider to be successful in running a guidebook business that is gold:

1. Have the physical fitness to perform this kind of work at the level you desire.

2. Willingness to take all of your personal and professional decisions required and be able to shoulder all responsibility upon your own shoulders.

3. Rely on your customers.

4. You must have the patience, maturity confidence in yourself and self-control to help you get through the moments when your business is not performing as well.

5. Begin by doing this part-time. It is common to require time to develop an enterprise into a viable business. Do not expect to begin your company and be able to have immediately enough cash to pay for every bill.

6. Basic self-management and organization skills for running a small-scale company.

7. The desire to learn and research about what you might not be aware of and what you require to know. The mere fact of doing business not enough.

8. You must be able to connect with people and love talking with them.

9. Possess a great personality, which people will admire.

10. Ability to handle rude people and personalities.

11. Make yourself a successful self-promoter salesperson and marketer. The business of selling is about marketing yourself and providing customer service.

12. Offer a quality service that is worthy of charging for.

13. Find an efficient and reliable transport system.

14. It is important to be able to transport customers to the guide places or meet them at the guide.

15. The most important resources are the equipment you need to demonstrate and use during your trips as a guide.

16. Get a clear idea of the kinds of services you can offer during your gold guide journey.

17. Be aware of how much you'll need to earn in order to make it happen for a long time.

18. Learn about the various areas that you can work and also the regulations you'll have to adhere to.

The Top Ten Reasons to be a Gold Guide

1. Being in your favorite sport doing, and being outside, and then being paid for it can be extremely rewarding.

2. Teaching people how to do better in something they truly love to do.

3. You can set your own schedule and set your own schedule. If you'd like to take the time off to take a holiday or to eat breakfast with your kids each morning, then you plan it off.

4. You can choose whom you will collaborate with and for whom you.

5. You will be in contact with people from all walks of life. People whom you meet

will be fascinating and you'll meet a lot of people.

6. You can choose which destinations you visit and make every visit a bit different by choosing to.

7. Your boss is you and you have the potential for being one of the best in your industry.

8. Small businesses can give you the chance to receive tax benefits.

9. Starting a guide company is very low in comparison to other small-scale firms. There is probably everything you need to begin your guide business and you can begin right away.

10. You are in charge of what you earn. You are able to earn money as an gold guide.

What kind of business should you start?

There are a variety of ways to start an enterprise that is small. Be clear about what you intend to accomplish from the beginning is crucial. Start by deciding on what kind legal form of entity would like to create - an LLC, Corporation, Partnership, Corporation, Partnership, LLC S-CORP or Sole Proprietor. What? This is a personal decision and should be taken care over by an expert. I suggest getting a consult with a qualified tax advisor or accountant who can help you solve these issues. You'll need examine many different aspects, including your finances, your tax situation and the tolerance to liability.

All this being said, the majority of guides that stand in their own right are sole proprietors. Being a sole proprietor also among the most cost-effective and easiest companies to establish for an unofficial guide. One of the positives of being a sole-proprietor include that your personal and corporate tax obligations are seen as one. If you are a sole proprietor, you are required to file one tax return at the close in the calendar year. Your personal and the business's income are treated as the same, as are any liabilities. Since you are sole proprietors the accounting process isn't as exact since the owner and the business are treated as one entity. However, it is vital to keep separate documents for your business. One disadvantage of being a sole proprietorship is that the entire debt tax, liabilities and taxes that you incur are in your own name. Therefore, I strongly recommend consulting an expert prior to deciding the type of business you would like to establish.

Chapter 3: Nuts And Bolts

Selecting an appropriate Business Name

It's a great time! The process of registering and choosing the name could be a bit difficult. The majority of names that you pick may already have been used, and you'll need to come up with a new idea when the name you want is already used. I suggest picking names that are gold-related so that it's clear what you're about and can help with web search results. Make sure you pick names that have a characters and has a certain zing. Be sure not to use the general name or nickname. I would highly suggest picking 10 names that have a connection to gold and asking your family and friends to rate the names they like from 1 to 10. Once they've chosen the names they like , you'll have a clear idea of the names that pop out at people they appreciate. Finding a name that appeals to a lot of people is an enormous benefit. If you've got an idea you like and you're ready to apply for registration and determine whether it's available. You'll need to contact your local city, town or state government offices to determine whether your name is available. If you are sole proprietor, it's much simpler since all you need is your own name.

Registration of Your New Business

Your business must been registered by the federal government, state and local authorities. The process of registering a business isn't difficult, but it requires the following documents to be filled in. It is necessary to contact every government agency to obtain the correct forms and filing fees, if they apply. You should find out the specific requirements that must be fulfilled in your area since each region is unique. Local and state governments provide assistance in the complete small-scale business setup procedure. The best resource to seek assistance within your community and in this procedure is your local government's Small Business Administration office. They're a fantastic source and can assist you in this process and guide you through the local laws.

The process of setting up a bank account and Merchant Account

The bank account you have is required. I suggest looking around for a bank that has an account for business that includes merchant account features. Merchant accounts allow the use of credit card payment options if you wish to. Examine all fees and benefits before signing up with one of them. You should find an institution that has low fees per transaction as well as low annual charges. Try to get the business checking account which provides a debit or credit card linked to the account to allow convenience. You'll need the card to purchase equipment and other things. I would recommend that your company's bank account is completely distinct with your private account. If the bank you have your personal banking requirements has a business account, then by all means, to ensure convenience you should use the same bank to manage your personal and business accounts.

The Tax Man

Yes, you'll have to pay the taxman and will be accountable for all state, local and federal taxes due. You are responsible to find out the taxes you're responsible for , as well. Take care of your tax obligations throughout the year so that you don't have to pay penalties. The good thing is that a guide company is not a lot of work with regard to the accounting documentation.

I would suggest keeping the track of your earnings and expenses throughout the year. Be sure to keep all your receipts and keep track of your mileage , as along with other miscellaneous expenses. It's really not that much of work, unless you make it more.

Tax preparation by yourself to run a guide-based company is feasible. Each person's financial situation is unique but. Deductions , as well as other things can influence your small business taxes and these guidelines can are subject to change. To know what's and what's not tax-deductible, you may be able to speak with a tax professional. A professional can assist you to organize your finances and offer tips on the best method to track your company's income and deductions, as well as expenses.

It's always beneficial to determine if there is tax-deductible for you, as deductions will lower the amount of tax that you must pay. Although tax rules can change regularly there are some things that could be eligible for deduction include: mileage and office equipment accounting software, books license fees, claim charges, office rent, telephone, etc.

Liability and Insurance

The requirements for insurance you face will be contingent on the nature of the business you want to establish and the method you choose to run your business. Insurance protects you from any losses or liabilities that may result from the harm or injury that your clients who use your products and services. Insurance could be a requirement due to beginning your business with the bank or by the government. Each local and state government differs in the way they require insurance. Certain states do not have any rules while others have strict rules and will check to ensure that you are insured. It is crucial to inquire with your state to find out the requirements for insurance. Contact your insurance agent and provide all the details about your company to figure out what you'll require for insurance.

If you don't have an insurance expert There are several ways to obtain and purchase insurance. You can contact the current homeowner's insurance agent or bank, accountant, or the local chamber of commerce to get suggestions of an insurance expert in your area.

Maintaining Records

I am of the opinion that no matter what type of business you choose to set up it is important to keep all the business documents separate from your personal financial records. Maintaining well-organized records can help in the year-end and month-end of tax time comes around. I recommend having a separate file folder for each type of record and keeping the files in a filing cabinet. The most fundamental records you must keep must include your the income and expenses as well as general ledgers for accounts as well as the essential listing below:

1. Payments

2. invoices

3. Equipment

4. Permits, licenses and entry cost

5. Complete contact details for the customer

6. Mileage

7. Insurance

8. Tax deductions that could be possible

9. Taxes

Formulas

If you are starting your own business, you'll need some basic forms in order to conduct business. You can design some of these forms on your own or purchase them from the local office supplies retailer.

One type of form you'll certainly require is a recreation activity waiver. This is a form you should have an attorney write it for you, and then have all of your customers sign. The waiver for recreational activities will be specific to your business guideline and require you to think about it with care as it could minimize the liability of your business in the event in the event of a dispute. In the end of the book you'll be able to look at the examples for these types of forms.

Here's a simple example of records and forms to keep:

1. Log of expenses and income

2. Mileage log

3. Maintenance log for equipment and equipment

4. Invoice book

5. Book of receipts

6. Customer feedback form/email

7. Recreational waiver

8. Thank you cards

Maps

Every great guide comes with maps of the region the guide works in so that they can ensure that his clients are on ground with gold bearing. The maps you've got are drawn by you, or purchased. It is important that the map performs the job of keeping accurate gold sampling and location records for you. Personally, I think it is important to be able to have top quality maps of the areas you work in. Once you've identified an area that you would like to work in , having it mapped and sampled is crucial. Make or purchase a huge area map. Then, small section maps. These maps will be where you will put the sampling information and then mark all the gold bearing regions you have located. The bigger maps will keep details such as access points, details of the structure as well as meet-up areas, flood note claim boarders, etc. The maps you save need to be maintained up to date constantly. You can purchase and find maps from numerous locations. Research and locate some places within your state or area with high-quality maps that help you. An excellent starting point would be the Bureau of Land Management. There are many other organizations that create maps of specific areas that you can purchase.

Permits and Licensing

Every state and local jurisdiction regions differ in terms of licenses and permits. Certain areas don't require permits or licenses. The locations you'll work in might require permits and licensing. It is important not to be a victim of ticketing or penalties or working on land that is not legally. Research and find out what you need to know prior to attempting. There are several locations across the nation which require licenses and permits for specific types of equipment employed. Certain types of land require permits for daily use. It is possible to find out what the requirements for licensing and permits are by contacting local office of the government as well as the Bureau of Land Management in the region. Be sure to adhere to all guidelines before you can enter the area.

Equipment

The equipment you'll require for guiding will be determined on the kind of guiding you wish to accomplish, the tools you are familiar with and what your client would like to be educated on. The possibilities are endless! It is possible to teach people about whatever they want to learn about.

When you are buying the equipment you need for your guided trips, be sure to do your research and purchase the best equipment. Get a few extras to cover the possibility of damages and losses. It is not a good idea to be in a situation in which you don't have the right equipment to go out with a group for a tour. Here's a basic list of the equipment you need:

1. Basic Equipment Basic equipment: Small shovel, gold pan buckets, snuffer bottle plastic vials magnet for sand and a Classifier 1/4-inch screen.

2. INTERMIDIATE TO ADVANCED EQUIPMENT Sluice, High Banker, suction dredge mini sluice size classification screens and metal detector.

3. tools for crevicing: vacuum magnets, spoon tools for crevicing, such as rockpick the hammer, chisels pry bar, paint brush.

It is crucial to care for and keep in good condition all your equipment. This is what will earn your money. Your concentration on detail and the care of your equipment must be the top priority. Make time on your calendar to look through your entire equipment to clean and repair any item that is required. If your equipment is damaged, repair it as soon as you can.

Make sure you have extra equipment, at the very least the essentials. Keep spare parts on hand whenever you embark on guided trips. The most important things to keep on hand include tie wire electric tape, duct tape, and an excellent multi-purpose tool.

Safety

In guiding safety, it is essential to contemplate and make plans for. Your and your client's security must be top of the list. I suggest, regardless of whether your state requires it or not, you be First Aid certified. Also, make sure you have at minimum a safety plan for your travels. Include phone numbers for emergency personnel located in your area, including fire, police and EMT. Have an emergency kit which is easily accessible throughout.

In telling your clients what they can expect from the day, also mention safety and where to find an emergency kit. If you are in an emergency situation, follow your plan as soon as possible.

Chapter 4: Systems And Processes

Systems

A note on Processes and Systems

I would suggest you have procedures and systems implemented to simplify your work and make life more manageable. In this section, I will go over some things you'll have to work on frequently to make everything simpler when it comes to working on your equipment, keeping your workplace, keeping tabs on your expenses and regularly assessing competitors, keeping track of self-promotion, putting it in your calendar, and accepting payments. For each of these things, it is essential to devise a repeatable process or system and then implement it at the appropriate times.

Set up your office or work Area

Establishing an office is not required however it is helpful in organizing all your business activities together and organized. It is essential to maintain a an organized system that includes all your paperwork and tools. There isn't much you need in your office. A phone, an answering machine, computer, desk and filing cabinet is all you'll require to run your business. If you don't have enough space in your house or office, you could make use of a garage, shed or even your car. As a minimum, be sure you have a portable organizer for keeping your forms and records in. It's easy to carry it with you, so that all the forms you need for a trip guide are in your bag and be in order.

Set Fees

Setting your fees is crucial. It is important to ensure that you have them set properly at the beginning, after which you should re-evaluate them each quarter. It is easy to do it yourself.

The fees of each person will differ due to the business costs vary widely. To determine your price determine the costs of conducting business, and then add the additional charge as your revenue. But, you must be sure to stay within a reasonable range of what other businesses charge, so that you don't go over or under over charging for your services. Find out what other guides are charging in your area, what they charge, how many clients they have at one time, and the amount each person is charged, take a look at all guides for fishing as well as the typical amount of days they offer their services in your region. Reviewing other guides can provide you with an idea of cost of market, the charges per person, charges for a group , and how much you can anticipate to make when you provide similar services.

Certain guides work only during summer, and must charge a higher rate per day in order to survive the winter months of slowness. Certain guides charge per full or half day, and others per hour. It is also possible to plan two half-day trips within the day to boost the potential of your earnings.

What is important is the amount you earn and whether you're receiving the resources you require to make it worthwhile for all the effort. You're doing this to earn money.

An example of setting up charges: You've calculated that all your business expenses will amount to $3,000 over the course of the entire year. Let's suppose you've calculated that you'd like to earn $37,000 while working part-time during the year. You believe you could manage to guide 100 excursions throughout the whole year. You can do two four-hour half-day trips every day (50 days in total).

Based on the study you've conducted, you've found the current prices for guided tours including fishing guides. Guides in the area cost between $300 and $500 for a 4-5 hour half-day excursion.

$40,000 per year x 100 tripsequals $400 per trip. A daily rate of $400 is the minimum amount that you must charge for each trip. Given that the typical rate is $300-$500 for a half-day guide, it is now clear that you could offer $400 per four-hour half-day or two half-day trips at $800 for the entire day. A $400/half day price set will make certain that you are charging the most competitive price for your service when compared with other guides.

The charges you make are solely yours! Don't use other guide charges and believe that you are capable of making it without knowing the amount of your personal bottom line must be for you to earn money. Don't cut yourself off or give the farm away or even break even. Be sure to keep records of your expenses in terms of profit, expenses and costs so you are aware of where you stand throughout the year. It's okay to adjust the fees you pay each year as it's a normal element of business. But, you don't need to set a low cost at the beginning, and later adjust them more because you've not completed your research or done the math. Set fair fees in the beginning, and remain conscient because customers will tell their friends about you and the price they paid for the service you provided. You should also read about earning more money in chapter 6.

Learning from the Competitors

I'm a firm believer in the fact that regardless of how hard I work to master something, I'll never master everything. There is always something to learn from others, and especially your competition. You must be the best and offer your customers with the best guidance service! Only way you can tell whether you're one of the best is to compare yourself with your competitors. You don't want to duplicate what they are doing Instead, you should discover ideas and improve them to make them more effective. Keep your eyes and ears open to anything they're doing that is wrong or something new. Find out what people say about your competition and the service they provide Then adjust your service to match.

Every year, you should set aside an opportunity to take a close look at your competition. Always look over everything your competitors do, how much they cost for their services, how professional they appear and what they are able to tell you about, how they travel and how they promote as well as the amount of work they have, what clients are saying, the services they have to offer, and what a typical day's trip is like and then attempt to make this better in any manner you could. If you come across something that doesn't work, then you should stay clear of it.

Don't be arrogant , or be you'll be viewed as a one-upmanship. Always keep in mind that you're an expert with direct concurrence. You must ensure that your business lasts for a while by keeping in mind that the playing field is never even. It's never hurt to be competitive. anyone, and everyone can have fun and make friends as we do what we love!

Self Promotion, Advertising Self Promotion

In the beginning it is essential to market your guide service. Self-promotion and advertising is an ongoing process that when done correctly, can pay dividends for the future of your small-scale company. Create a plan of what you'd like to accomplish and then tweak it over time until you can ensure it functions like an efficient machine. Once you've figured it out, repeat it frequently to ensure that clients keep coming back.

There are numerous ways to promote your company. If you're doing this part-time until you are enough of a business to become full-time, I suggest that you do not spend too much on advertising. Spend lots of time disseminating the message and distributing your information to as many public locations and individuals as you can through cards, flyers, and telling your story.

In spreading information, your primary and most crucial aspect of self-promotion is to be very enthusiastic about the work you're doing. Your enthusiasm and enthusiasm display will inspire people to join your business and make them want to use your services. When you introduce yourself and talk about your business ensure that you're happy to provide a wonderful service at a reasonable price.

People are drawn to those who are full of passion and enthusiasm for the things they enjoy doing. Additionally, you must satisfy people's desires by teaching them how to do something they'd like to learn and to show them what they are enthusiastic about. In helping others, they'll be thankful and will share your name to their family, friends and anyone else with similar interests. We all are aware that word-of-mouth can be an best and most powerful way to advertise. To begin make sure you are enthusiastic and share your story to the maximum number of people you can. Fill their needs in relation with your service as a guide.

In the beginning spreading information about your business by sharing it with your family and friends. You will then need to make your way throughout the town, telling people of what you can provide. Make sure you focus the efforts you make on regions that attract tourists to your region. Connect with other owners of businesses in your area that cater to tourists and provide services to tourists. I would suggest establishing relations with hotel concierges as well as the owners of hotels. Offer them your tickets and brochures. If you're out and about meeting new people, take your time and become acquainted with them. Take note of something personal about each person you meet. Keep specific notes about each person particularly your main clients. This will be helpful the next time you have to meet. Establishing relationships that last can be a very valuable resource and can allow you to sustain your company for long period of time.

Here are some ideas to aid in spreading your message about your service and help you get the ball rolling for a brand new business:

1. Web Page: With few pages like About you, Services offered and contact information and what you can expect to pay costs, testimonials, and questions and answers

2. Motels and Hotels

3. Historic sites

4. Shops for gifts

5. Stores for prospector supply

6. Metal Detector stores

7. Boutiques selling gifts

8. Cards and Flyers

9. Connecting with other guides, fishing and Rafting

10. Rec Centers

11. Participating in outdoor groups or with programs

12. Word of Mouth

13. Bulletin boards for public announcements

14. All small businesses located along the river or the area that you operate in

15. State travel brochures

16. Newspaper article on you

17. Stores selling sporting goods

18. Travel guides

19. Blogging

20. YouTube videos

21. Facebook and other social media

How to plan and schedule Clients

Before scheduling even one client, you'll require a new calendar, or a different system to track all guide days scheduled. It is essential to keep track of your schedule and time. It's essential and will in saving time and money. Believe it or it's not. The ability to track your hours will allow you to be more productive when you're at a loss. It is important to determine the amount of time off you require for in the coming six months, and then you can begin scheduling clients. I keep an electronic calendar on my phone to mark every day I'll be working for the next six months. Yes, I said six months! Clients will contact you and arrange their vacations according to your schedule several months ahead. You're a professional today and must make plans far in advance. The best times to mark on your calendar for the coming six months are holidays that you celebrate such as birthdays, holidays holiday days, anniversaries, vacation days or days to rest and other days that you think you should leave to keep your life enjoyable and satisfying. Once your calendar is completed and a call comes in you, you'll be able to plan your schedule with no unexpected conflicts.

An organized calendar allows you to know exactly what's open or scheduled. Request from the customer whether they would like an hour-long guide trip or a full day guide tour and record it on your calendar. Both you and the customer have a date that you have both agreed on.

Making Payments

It is up to you to decide the best time and method to accept payments. It is important to make accepting payments as simple for yourself as possible , and offer the customer several choices. Offer the option of paying with credit or cash. It is possible to offer checks , however I'm not a fan of taking checks due to the issues associated with having insufficient funds.

Some prefer to pay in one go the week prior to their trip, and that is okay with me. In this case I'll make the check so that I can assure that it will be cleared prior to our tour. Payments can also be accepted via PayPal and, if you've got an account for merchants that accepts credit cards easily.

My clients typically make a $100 deposit when they book the trip. The balance is due on the day prior to the trip, prior to starting. Deposits will ensure there are only serious clients who are able to book your time. A majority of uninterested customers will not pay a deposit in order to secure the date. Making this deposit of $100 sets the day as a fact for the client. I know that I have a high-paying client scheduled to meet me and I don't need to think about filling the day now.

Refund and cancellation policy

I suggest that each guide includes a refund and cancellation policy. If you have an established policy, it stops those who want to cancel and allows you to keep some of the income you lost.

EXAMPLE POLICIES:

Cancellation Policy: If the trip is cancelled within 30 days of advance, the total amount will be refunded, or can be used to fund a subsequent trip. In the event that the journey is cancelled within 3 to 30 days and 30 days, 50 percent of the deposit will be returned, or the total deposit is able to be used for the next trip. If the trip is canceled within three days of the departure date the deposit will not be refunded.

Refunds: In the event that for any reason, we have be able to cancel our trip for any reason, all deposit will be reimbursed or used to pay for another trip.

It is recommended that you do not postpone or cancel your trip.

What Clients Need To Bring To The Guide The Trip

Be sure to inform your customers of what they need to bring on the tour. Make sure to post the details on your website and any confirmation email you can send.

Below is a list of items I suggest clients bring to ensure their safety and security:

1. Camera

2. Hat

3. Ziploc bags (to protect valuables from water - such as cellphones)

4. Sunscreen

5. Long sleeved or jacket

6. Rubber gloves

7. Waders for your chest or hips for those who plan to be in the water and not near the shore

8. Bug spray

9. Food Drinks, snacks, and food It is recommended to always bring drinking water in bottles and avoid alcohol to be consumed on the trip.

10. Equipment This guide will include all the equipment (such as buckets, pans and classifier screens, snuffer bottles and shovels, sluice boxes, etc.)) which will be utilized for the journey. If the customer has an item of particular importance they wish to use, it's acceptable, however there may be an additional cost.

Chapter 5: The Client

How to screen your clients

Yes, I said screen your clients! It may sound strange but it's essential, particularly in the guide industry. Some clients are not worthy of your time! Some customers can cause you to lose time, money, and cause an enormous headache taking a guided tour. I believe that 1-2 percent of clients you encounter might not be worth the time when you go on a guided tour.

You'll need to devise the method of screening your clients to meet your personal preferences. My screening process begins with my first meeting with my potential clients. I start by asking them numerous questions about what they are looking for and what they expect. For instance, if you talk to clients who keep asking to be put on the spot where they could discover gold nuggets and one ounce of gold after they're done. This is a naive belief that a few people actually have , and should be explained upfront. If, after explaining the reasonable expectations, the client persists in believing that they will discover a gold nugget that will satisfy them, then you may want to reconsider their status as a customer.

Additionally, you should learn about the expectations of your prospective clients and needs for the trip and their physical and health capabilities too. Physically, it's the longest and most strenuous few and sometimes more. For certain people who are physically challenged, this might not be an option and both you and them should be aware of their physical demands and limitations. Also, I might alter my route if I am aware that the person isn't able or unwilling to trek 2 miles in an unforgiving canyon. It's fine to inform anyone that you think this arrangement won't be successful, and they might discover a trip that is more appropriate to meet their needs with the guide of a different company.

As a professional, you don't want to place yourself in a situation in which you're unable to do your job and trying to go above and beyond for a client to make a profit. You'll discover your own opinions about the people you would like to work with and also the few you do not want to collaborate with after you have been guiding for a time. Keep a record of the things you enjoy and dislike throughout the process. It can help you to have more fun and be with people you enjoy working with.

Making Your First Few Clients

It's often difficult to locate clients at first. Some suggestions for getting your first clients could include:

1. Have friends and family members ask questions.

2. Neighbors

3. Church-going friends

4. People at work

5. Distribute flyer

Additionally, you will need others to assist in establishing your abilities in the beginning to test your service and your trip. You might want to provide some guide tours free of charge until you feel at ease that you've planned and planned a great day excursion.

After you've taken several people to the beach and worked out your problems, then you can ask them to write reviews about your services. Be sure to publish positive reviews on your site so that future customers can see.

Customer Service

Customer service can be the difference between success and failure in this highly interconnected business. The best guidebooks for money are ones who are enjoyable to hang out with and possess excellent personalities.

If you're really great with customers, the word will be spread about your name as well as your base of clients will expand. But, on the other aspect, if you're poor with customers, your reputation will also spread, and your bank accounts will reflect it, and your client base will decrease. A well-paying, valuable customer an enjoyable experience that is top-quality and outstanding customer service can take away the competition.

Here's a list my most important customer service tips I would recommend using as many as you can:

1. Everyone you meet has being treated respectfully and treated as an ideal potential customer.

2. Be honest and clear with your customers regarding your services and what you don't offer, so that your customers know what they can expect. It will help them comprehend and plan for their journey.

3. Your clients should be addressed by their first names frequently. Note their full names on your monthly calendar and ensure that you address them by name at

the beginning of your meeting for their tour guide.

4. Let people know that they feel important. Pay attention when they complete an activity correctly and look for ways to show them appreciation by giving them a heartfelt praise.

5. Let people know what they need! Make sure you are an excellent listener and determining what their requirements are. Once you have a clear understanding of their needs you can offer them top service and exceed their expectations.

6. If you've made an error, be sorry and correct the problem in the quickest time you could. Make sure that the issue is resolved for the client.

7. Provide your customers with the best personal experience that you can. You must ensure that your customer service is superior to that of your competitors.

8. Please be respectful and thank you. Send your customer a message or a thank you note card, or a thank you note within

a few days of their trip guide. Let them know that you were pleased to meet them and enjoyed the company of their guests.

9. Your friends and family members should be treated exactly the way you would treat your customers.

10. Engage with your customers as if they are your best friends. Don't speak to them as if you're trying to sell your services (or a package of sale since it turns customers off.

11. A good caller message to your mobile. Make sure to return calls as quickly as you can.

The customer service extends beyond the guidebook trip and can last several years for some customers.

It is also important to hear feedback from your customers following their experiences. Make your customers part of your improvement process by giving you their thoughts about what they liked or did not like. Do not be afraid to change what didn't work and building on what worked.

It is important to understand that a gold prospecting guide doesn't mean being out every day panning. It's showing others how to prospect for and locate gold on their own. It's extremely satisfying to rise every day and teach others about what you enjoy doing. Guiding is a business for people. It's a sharing information and expertise.

Chapter 6: The Whole Picture

Start-Up Checklist

Take a look at the list and ensure that you've got things in order prior to beginning.

1. Name of business to be registered.

2. Set up a bank account.

3. Merchant account set up.

4. Register your company.

5. Web page has been set up.

6. You've got flyers and cards designed.

7. There are all maps that you will require.

8. All the forms are available, including the waiver of liability for recreational use.

9. You've got an office installed.

10. You have all the tools you require.

11. You've picked an area out, mapped , and taken samples of it.

12. Your calendar is set to the rest of the year, and you have the time off.

13. You are aware of exactly what, when , and where you'll be on your guided tour.

14. You are aware of how to make payments and which forms of payment you'll accept.

15. You've been able to spread the word around the town about your guide service and the services you provide.

16. You've got a couple of potential clients for testing.

The Guided Trip 101

After you've had everything done and you're now ready to start coaching, you're ready to sign up your first clients who pay. Create a half-day plan and a full-day plan, and record it on paper. Include all the information of your tour you'll need to keep in mind from beginning to finish. Be sure to keep your guide's itinerary punctual from start to end.

Here's an example of a basic list of things to be done when you first start your day as a guide:

1. You've got the day guide laid out from beginning to the end.

2. You are aware of when and where are you instructing the client.

3. Client has already arranged for a meeting place and time.

4. You will meet with you with the client(s). Make any necessary introductions.

5. Make the remainder of payment. The client has to be fully paid prior to the time you begin.

6. Let your clients sign the recreation liability waiver.

7. Review your schedule to discuss the plan with clients.

8. Go over safety.

9. Ask any question you want to.

10. Discuss the region and its history, the basics of prospecting, the movement of gold in rivers and other rivers, etc. Ask any question you have.

11. Equipment is given out, along with an explanation of how to use it is included.

12. Begin by heading towards the guide's location.

13. Learn what the student is hoping to learn with enthusiasm, a smile and enthusiasm.

14. Visit all of your places to sample, pan, and sample and identify the most important gold-bearing areas throughout the process.

15. Between one and thirty minutes before the trip ends, You announce the tour is coming to an end and that you will all be returning.

16. Inform the client that you would like to answer any questions they may have or assist with techniques or technique. or any

other issues that clients may wish to address before the end of the day.

17. Retract your head.

18. Collect all equipment.

19. Thank them and congratulate them for the job they did well.

20. Offer additional guide services in the future as well as referral discount.

21. Clean up and organize. Get ready for your next clients or head to your home.

22. Make a deposit in a bank in case you need to do so while you are on the way to home.

23. Clean the equipment and then reorganize it.

24. Get ready for the following day's tour.

25. Answer and return all calls and emails.

26. Make sure you take care of paperwork for business, such as making forms, keeping track of the expense and income,

miles samples, and taking charge of maps should the are required.

An example of the Guide Trip from beginning to end

When you have decided on the kind of guiding you would like to offer and where to go you'd like to go, then you'll design half - and full-day trips. I design all my trips in accordance with the requirements and wishes of the client. Every step of the learning process will differ in the duration of time required to teach and is determined by the abilities of the client and the number of clients within the group you're working with on the day.

This is an example of how a typical full day guide tour looks like for a handful of beginner panners looking to learn the basics of the trade and how much time I'm spending on each one of these activities:

1. The day prior to a trip, I make a call to arrange a time to meet together with my client. I then clean and then pack

everything I have brought along to take on the next day's adventure.

2. 8 am: In the morning before the trip, I meet with the client at a location that is convenient for them. We after which we make introductions, present an overview of the schedule for the day taking payment, going through safety and inquire about any health or physical issues I need to be aware of prior to when we begin. I will then ask them to sign an agreement to indemnify. Inform the clients that they are keeping any gold they can find and then give them a new plastic vial to every one.

3. 8:15 am: begin my guided tour, going through the equipment we'll be using and their purpose. Distribute each participant's equipment.

4. 9:00 am: I review the basics of finding gold, how it is transported through rivers in the first place. I also explain what things to search for. I provide examples on paper as well as by walking alongside the river.

5. 10:00 am: we travel to the river to begin the basics of the panning technique procedure. The student must be confident that the technique is correct and pan before moving to the next step. Have a handful of balls with the students to use for practice.

6. at 1:30 pm: I begin walking them along the river, and then demonstrate the "How To Find Gold Quickly" system for locating gold, and to take samples as I go.

7. 3:00 pm: Inform clients that their that their trip is coming to an end. Offer to guide them for an additional few hours and also teach additional hours in case they want the challenge. Continue to guide until the agreed-upon duration or return to the vehicles.

8. 4:00 pm: The trip is over! We want to thank for the trip! Thank the client for their hard day's work. Offer discounted continuing education guides for excursions.

If you're doing an excursion for a day you can create a condensed variant of what you have described above. Be aware that to earn an extra income, you can plan two half-day trips per day , so you need to be aware of your schedule because you don't want to begin your second session later.

Chapter 7: What To Consider About

The Biggest errors that Guides Make

If I had known then what I know now starting out was a lot more simple. Below are a list of errors that are frequently made. By avoiding these mistakes it will benefit you significantly:

1. Do not provide top-quality quality customer care and service! It will end your business.

2. Not pricing your service properly. Be sure that your pricing is comparable to others, and ensure that you don't cut too much in order to aid someone else. Don't give up your time to help others!

3. You may think that because you went through town selling your services and yourself during two months, you'll never need to repeat it. You should always be selling your services and yourself even when you're not working. Make sure you are filling your monthly calendar with work!

4. Do not schedule your clients and yourself correctly. Always keep track of your calendar's details.

5. Refusing to return calls and not returning calls to prospective clients as quickly as you can. Rapid, transparent, and prompt communication is essential. You don't want to lose customers due to the fact that they contacted another person after it took your time to reply to their call.

6. Spending money on advertisements every time that isn't working. Be aware of what works and what isn't.

7. Setting the wrong expectations and limits with clients.

8. Awakening with a negative attitude and taking it to your guided excursion.

9. Doing not take care of your appearance.

10. Do not maintain and clean your equipment. Make sure you have clean, clean, working equipment.

11. Not getting ready for an excursion.

12. Do not accept multiple forms of payment. Certain customers only pay using credit cards, while others prefer to pay via the internet. There is a risk of losing clients in the event that you do not accept several types of payment.

13. Insufficiently proactive and trying to fill your calendar ahead of time. If you notice that you have an a calendar that is empty it's the perfect time to get started promoting your business by contacting all of your contacts, sending emails with solicitations and giving out flyers.

14. Don't spend your time answering messages from similar individuals repeatedly, who are simply looking for information. There are individuals try to force you to answer their queries without paying to share your knowledge. Respect them, be polite inform them of the you can offer in terms of services as well as how you could assist them with their requirements, and then gently inform them that you're busy right now.

Time-saving and Money Saving Ideas

Over the years , I've discovered a variety of ideas that help me save the time as well as money. Here are some ideas I have to think about:

1. Your customers are helping you to sample vast areas! This is huge and could make you money if they assist you locate a great gold streak. Also, throughout the journey or at the close of each day, make a note of the results.

2. Learn to set up and use your equipment properly. If you are unable to comprehend the new piece of equipment or aren't able to set it up correctly it could cost you gold in the process. Be aware of the equipment you are using and be sure to ask lots of questions prior to purchasing it.

3. Set your goals for the day to week, month, or longer. It will save you lots of hassles in many different areas. My grandfather wisely advised, "If you fail to plan, then you plan to fail". Plan

everything out and take note of the details that are in between as much as you can.

4. Include spare parts in your car and carry a first aid kit as well as emergency numbers handy every time you go out with clients.

5. Each quarter, you should calculate on your total income and expenses, and then adjust your costs and prices accordingly.

6. Don't stop promoting yourself, even when you're busy or not leading. If you're out for dinner and someone asks you about your services, inform them about your services. Make sure to spread the word.

Earning Even More Money

If you experience low times during certain seasons or you want to earn more money, here are couple of good options:

1. Group lessons: Choose some days in the months off your schedule, and then set the date! The months of spring and autumn are ideal time to schedule this.

Make sure to announce all over the world that you're offering an in-group lesson at the lowest cost. After the lesson has been completed you can offer guided tours and consults to show advanced techniques or equipment. Keep your calendar handy!

2. Consultations: You could provide assistance to people in learning how to use the equipment that they recently purchased or plan to purchase.

3. Teaching in local schools Some community colleges let you teach or organize activities for their students.

4. Emails that are semi-annual and offer discounts on your services If you're looking to increase the amount of work, an email promotion is a fantastic way to add more clients to the calendar.

5. The Last Minute Discounts If you have open dates on your calendar which need to be filled hand mailers or flyers and phone people and take your time making rounds across town. Inform everyone that you're offering a discount on last minute

calendar openings. Offer clients referral discounts.

6. Paydirt! Paydirt can be collected and sold via a website and in local stores.

7. You can sell the tools you have used. Be a dealer for the brand you prefer and employ. Customers are always impressed by my equipment and would like the same. Learn to market your equipment to your customers during guided tours in a manner that demonstrates how it functions and how wonderful it is.

A Note About Your Sponsors

Sponsors are companies, dealers and representatives of businesses that wish to let you use to test and sell, promote and market their products in exchange of goods or cash. Sponsorships can be beneficial for your company's financial success! It is possible to have an organization that sponsors almost everything including clothing, vehicles, equipment or gear including tires, batteries and so on. Guides love sponsors since they can help to pay for gear and lowers the cost of overhead. Equipment can be purchased for half the price or even more, then market it to your clients. Some sponsors don't require anything from you they simply give you the product.

Sponsorship can be a hassle! Depending on the sponsors' requirements and the sponsor's it could result in an obligation you do not wish to make. Before working with a sponsor , think about what you want and can be in a position to give each one another. If you plan to promote and sell the product or service, what do you expect from the deal? Do you think it is worth it?

If you are considering working with a partner, be sure that you research and show that it is a good fit for you and your company first. A good fit is when you like the product or service they provide and that you have used the product, and will do whatever they want in return for the product or cash you receive. They should also be a fan of you and willing to offer you a product or cash in the event that you endorse, showcase and use their products. It can be a win-win situation if it helps you save money on the cost of overhead for products every year and allows them to promote their product.

Make a calculation of the cost or how long of the time commitment it will take before you sign up. Don't sign up with a sponsor if don't believe in the products or simply be an organization as a sponsor. Sponsors are looking to partner with guides and winners who have a good reputation and they are able to help in spreading the word about their product.

Chapter 8: Questions And Responses

1. How do you deal with an angry customer? Answer: First, never lose your cool with a client. Be courteous, respectful and try to resolve the issues with the timeframe you can. Let them vent out their frustrations without talking too much and simply listening. Your aim is to communicate with your client in a manner that will allow them to keep their dignity, and to leave with respect for your respect. You may find that they've been through a rough day, and you'd like to get to get their recommendations or every other company they could refer to you.

2. How can I manage being overwhelmed? Answer Make sure you have your calendar handy and with you constantly. Never over commit yourself! You must are able to schedule your time off in the calendar six months in advance. Make sure you plan your entire day free with loved ones , and then grant yourself set days off each week. Time off can be valuable and should be cherished and not squandered. If you can

set your hourly rates correctly, you will never feel like you must work more than eight hours per week.

3. What can I do to stop customers due to my busy schedule? Remind them that you aren't capable of taking them out and then recommend that they maintain their email addresses as well as all contact details. Inform them that, if you need to cancel your appointment that you will inform them. Additionally, suggest offering them a discount when they make an appointment an appointment later.

You must call these clients back and make sure you have your calendar ready when you call them back. Make this a habit for each customer who wants to be a part of your trip. It's true that being at the pace you like and earning the amount you desire is the best situation to be in. I refer to this as the sweet spot, and I recommend to stay in this spot all the time. If you have to, you may increase your hourly rate to the level of fair market value due to the fact that you are highly sought-after. DON'T GET GREEDY!!! Customers want quality service at a reasonable cost.

4. What should I do when I'm running slow and I don't have clients scheduled on my schedule? Answer The answer is that you should be following an effective self-promotion plan and working it prior to the time you get to the point where you aren't working. Learn more about self-promotion in chapter 3. The best option you can do is to hit the streets and inform anyone you know about your service. It is necessary to send out emails or make phone calls and visit with store owners as well as any other people in the area regarding your services.

5. I'm always working but don't have anything to show for it. What can I do? Answer: You should keep an eye on your time and the total cost. It is possible that you are not priced correctly and could be wasting your time. It is essential to track your charges every quarter. Review your fees and time, then make sure they are set

correctly so you can earn the cash you require.

6. What should I do if I see other people at the same location to take participants on a guide excursion? Answer: Be professional and considerate of other river users. It's not a good idea for you to get angry and become angry. Most likely, the other people don't know the full extent of their actions in the first place and won't clean off your mess. There's something to suit all. Also , a reliable guide will provide a variety of locations to work from within the areas he has designated as his guide. There's nothing to be angry about unless there are some locals who are constantly chasing you and working at your location. If you do not have a claim on every spot you'll have to contend with those who are who are on your spot occasionally.

7. What happens if the weather is too bad for me to go out, or if I'm sick, or a client is unable to go out? Answer: Go out, rain or shine, regardless of the weather except if the person isn't keen to go out or you're

too sick to go. Connect with other guides and rely on them to assist you and help them in the event that they require it. If the customer does not wish to travel, or you're not able to to go, then offer a complete refund of the deposit as well as any other payments they've made to the trip. Request that they reschedule the trip for a future date. If it's your fault that the trip was not able to go ahead, as a supplement your full reimbursement, consider offering an additional discount on a subsequent trip.

Do not keep a payment or deposit from the payment of a client in the event that they didn't receive any assistance from you. They will be able to remember that you were helpful to them, and this will pay off in the future.

8. What happens if my clients are unable to find gold during their trip? Answer: Inform clients at the beginning of their trip you cannot guarantee that gold can be found. Be honest with your client as well as remind them there's no assurance.

Sometimes, it happens. The customer will typically recognize that it happens. Don't promise or assure the customer that they'll find gold, unless you can. It's okay to promise gold on every trip if you are in a gold bearing zone and there is gold flour in every pan.

After all the above mentioned it is time to complete your part and place clients on gold as often as you can. A solid guideline can ensure that each client walks away with a small amount of gold. It isn't a good idea to have clients to leave empty handed.

Chapter 9: The Inactive Look For

Silver And Gold

Therefore, I could of course on classifieds sites to advertise e.g. private collectors are looking for silver/gold jewelry and gold/silver coinage will pay top cash prices

Here's an example of newspaper ads where classified ads are displayed.

Another option is to distribute flyers and then hang them in the supermarket such as. Giving out business cards to cars is

another method to reach out to those who might be interested. It is a great idea to leave your business cards and flyers in retirement residences. Make sure to offer a complimentary evaluation service , and the chance to see the gold or silver jewelry or coins on the spot. It is essential to build trust because trust can make you a good business.

I'm in a small town of 140,000 residents. I went to the grocery store on a recent Saturday , and noticed several signs that read, "Money for Gold and Silver today at the Sheraton Hotel" After I left the market, I drove over to Sheraton Hotel. Sheraton Hotel to see what was happening there.

The gentleman was sitting in one meeting room the lobby, equipped with a table, scales and a calculator, as well as cash register. He was chatting with a lady as two others waited for their turn. I was not wanting interrupt him however the woman who runs the reception is a good friend of mine and I wanted to know what he was up to. She informed me that he saw a steady flow of clients throughout the day. Imagine if each customer bought just one ounce of gold and could have seen around 20-30 people on the day. This is a lot of money.

The gold market has been always a lucrative business, but it's not the same as it was before the year 2008 when prices for gold increased after the economic downturn began. This and other facts teamed up to force many people from their homes and sell their gold. They either need cash or they desire to profit from today's price-tag.

None of these methods is as effective as a cheap, simple marketing method

Do you love travelling? Some time ago, I had the pleasure of meeting an elderly man and his wife, who travel across the country in the camper van. They buy gold in each town they visit. They even put a notice on the back of their motor home stating that they are purchasing gold and that people stop by from their motor home.

Gold Parties are a booming business.

One interesting option is to hold gold-themed celebrations. We've all heard about private parties at home to purchase Tupperware and other cooking tools. At these events the guests spend money. when they attend an event that is gold and earn money! If you can invite 10 people to attend a Gold Party, you can purchase multiple tons of gold and make a profits of 45 percent on each ounce. The hostess who hosts the event will receive 10% of the money you make. It's an opportunity that is win-win for everyone involved.

Based on the method you decide to go depending on your preference, there are plenty of options to locate sellers and buyers, so let's look at the most crucial issue of buying silver and gold from private individuals.

7. How do you safeguard yourself from counterfeiting?

The majority of people buy their silver and gold through dealers instead of private dealers because they're scared of counterfeits. However, if you are aware of certain things and perform the tests of precious metals on your own, you won't need to worry about counterfeits.

In the case of silver and gold jewelry, there are hallmarks, which are marks that identify the content of silver and gold. They can, however, be fake, and it is recommended to always conduct tests. More details will be provided on this in the future.

In the majority of countries, there are regulations that are legally enforceable for the hallmarking of precious metals. However, these regulations differ from one country to the next.

Anyone who deals with the subject of precious metals can't avoid the terms carat and hallmark. They typically indicate the quality of silver or gold and can be used to quickly calculate the weight of the metal.

What are the characteristics?

When the term 'punch' is spoken or heard in any way, a clear distinction needs to be drawn in between the two definitions. A hallmark is generally the stamp that identifies the quality of silver or gold on an item of jewelry or ingot.

But, the instrument with the gold and silver stamp is engraved into the precious metal is known as the hallmark. The tool used is a hammer which is placed on the precious metal and then leaves the embossing area with the force of a blow.

According to German law it is not required to mark the quality of silver and gold in the metal as an initial. The hallmarking process does not need to have to be done but it is nevertheless recommended since the quality is evident in a glance. Another check for quality is not required during the initial step.

If the silver or gold is hallmarked it is required to be done in line to the law governing the quality of silver and gold goods (FeinGehG).

In addition, this is a way to regulate the way that

The gold that is stamped must have an purity of 585 parts per 1,000 or greater and

Punched silver should be of a minimum quality of 800 or more millionths of millimeter.

This guarantees that a certain value can be derived from made-to-order precious metals.

Furthermore, a maximum tolerance of error of 10 parts for million is set for the hallmark stamped. This is equivalent to a variation of one percent between the hallmark and actual quality gold.

The distinction between hallmark and carat

The words hallmark and carat are in reality just two different measurement units to determine the quality of silver and gold.

The number on the hallmark signifies the high quality of the item in relation to the overall weight. The calculation is made in millimeters. For gold jewelry this means that a mark with the number 833 represents the presence of fine gold at 833/3000ths. This is an extremely fine gold content of 83.3 percent for this particular item of jewellery.

In based on weight and the weight, the exact amount of gold can be determined. If the jewelry in the example above weighs 100g, then the exact gold amount is 83.3g.

As previously mentioned the term "carats" is another way of describing the high quality of silver and gold. In the next list, you will find the carat numbers that belong to the hallmarks that are associated with them.

Gold Karat Information Chart					
Karat Gold	Parts Gold	Percentage Gold	Decimal Part Gold	Normal European Stamping	Normal American Stamping
9 kt	9 in 24	37.50 %	.3750	375	-
10 kt	10 in 24	41.67 %	.4167	416	10k, 10KP
12 kt	12 in 24	50 %	.5000	500	-
14 kt	14 in 24	58.33 %	.5833	582 or 585	14k, 14KP
18 kt	18 in 24	75 %	.7500	750	18k, 18KP
22 kt	22 in 24	91.67 %	.9167	917	-
24 kt	24 in 25	99.99 %	.9583	999 or 99999	24k

Alloys - What's that?

If the gold jewelry piece is not made up of 99.9 percent gold, it is referred to as an alloy. They are composed of various metals that are joined in the process of melting.

Although pure silver or gold is certainly the most valuable metal, both in industry and jewelry manufacturing the alloys of silver or gold are used primarily. This is due to the fact that pure precious metals tend to be too soft.

For instance, necklaces made of gold with the purity of 99.9 percent would easily break. This is why silver and gold are enhanced by other metals. They are more durable and, consequently, more resistant. Additionally, they can be employed to alter colors of the precious metals. This allows for the creation of jewelry that is much less expensive.

Carats and hallmarks are important in the world of trade

Hallmarks are crucial to trade around the world. By indicating the quality of the coins, for instance, bars made of silver or gold can be traded with greater ease. If the bars are sold through reputable providers It is not necessary to test the quality of these coins.

In the context of international trade, however gold should be treated as a distinct instance. The term "gold" in different nations is based on its purity. The fineness of gold is a factor in Great Britain and Switzerland, gold is described as gold with the hallmark of 325 (9 carats). In the Netherlands there is even called the gold of 585 (14 carats). The Middle East and Asia it must be remembered that it is rare to find pure gold that has a total content of greater than 22 carats (91.6 percent) is interesting.

To make it easy to check, you'll need an magnifying glass, an adsorber and scale. This will allow you to make basic tests.

Pure gold can be described as diamagnetic. This means that it repels when you apply a magnet against it.

Inspection of the visual;

Particularly in the case of fake coinage of gold, mintage usually is different from the real thing in minor specifics. Condition If you have an original available for comparability, or you have a picture from the actual.

Dimension and weight:

For all gold coins that are known the weight, diameter, and thickness are known for all coins. The actual copy of the coin may be significantly different from these. Needs: A very fine measurement device (caliper or balance). For tungsten-based counterfeits there are no differences to determine for the layperson. Gold dummies made from the tungsten alloy are nearly as heavy as gold.

Test of water:

The water test can be used to determine the particular gravity of a sample. All you require is a glass that has been filled with water. The sample is fully submerged. The particular weight of the specimen could be measured from the displacement of water.

This will require a half-full bottle of water as well as a fine scale that has a digital display. This test of water, referred to as submersible weighing an invention of Archimedes. According to him, the weight of an object is determined by the amount of water displaced. Once you have the precise weight, you can also determine the kind of material it should be made of.

And here's how the gold-water test functions:

DENSITY TABLE

a.)Research and note the particular weight of your sample. Every Gold alloy comes with its own unique gravity. These numbers are well-known. We require the values to make the adjustment later.

For instance, if we own an emerald ring made of gold with 585 hallmarking, the exact weight will be between 13.02 g/cm3 to 13.64 g/cm3, depending upon the alloy type. The following details on the specific gravity of specific gold alloys is:

b.)Weigh the specimen using an accurate balance and note the results (=weight by grams) c.)Place the glass on a balance and set tare to "0

d.)Immerse the sample into the glass with string. The water should not flow over.

e.)Read the weight change (glass + sample immersed)

f.)Calculation The weight of sample multiplied by weight of the glass with the sample submerged.

g.)The results must be in line with the values that are known for the exact weight. If not, it's fake gold.

Conclusion: right in the theory. However, in practice you must be precise in order to obtain a strong outcome. It is not suitable for gold fakes made from tungsten, or jewelry pieces that are hollow or contain air inclusions.

It's best to are familiar with the tone. To accomplish this, rub the gold coin with a different metal object. The gold "sings" while fakes sound dull.

Bars' characteristics:

Genuine gold bars include embossed stamps from the manufacturer, that includes information about the metal used to make them along with the grade and weight. In the event that any of those data do not exist, then doubts should be addressed. For larger bars, the serial number will be added.

Chapter 10: What Is The Best Time To Purchase Gold, And What Is The Best Time To Do So? I Prefer To Purchase Silver?

The answer to this question is in the easiest manner by looking at the gold silver ratio, also known as the gold ratio.

The ratio between silver and gold indicates how many troy ounces silver are required to purchase one troy the ounce of gold. It is a value or price ratio between both precious metals. The more valuable the ratio between silver and gold, the less silver value is compared to gold. This ratio can be calculated simple by dividing the value of gold of US dollars for each troy ounce by cost of silver per troy ounce in US dollars for each troy an ounce. The ratio between silver and gold is utilized by investors to gauge to gauge the development of the prices of both precious metals. In order to understand the significance of this ratio the ratio, a comparison to previous ratios is essential.

From the beginning of the 1800s, the silver:gold ratio has varied in between 1:10-1:100 reaching a previously historic low 1:7 ratio in the Middle Ages. When silver was more expensive than before at the beginning in the fifteenth century the precious metal started to decrease in value, which was evident in a rising ratio. As in the nineteenth century the silver standard developed to become an industrial metal. The silver standard that had been in place in Germany from Charlemagne was changed to an gold standard, which was introduced in 1871. The ratio of gold and silver was 1:15 for many years, it increased to 1:30 by 1873, and remained that way until the beginning of 20th century.

In the 1970s The Hunt brothers made sure that silver's price was able to climb to around USD 50 an one ounce through the intention of buying silver contracts through the commodity exchanges. But the speculation bubble this caused burst and the value of silver dropped to less than $5 per troy one ounce. The massive price drop was also evident in the ratio between silver and gold, therefore, it was necessary to purchase 100 troy troy ounces of white metal to acquire 1 troy ounce gold. Since 2001 the ratio has generally been between 1:50 and 170 but with one exception: 2011 in which the ratio between price of silver and gold decreased to 1:34.

Here's a graph of the ratio of silver to gold:

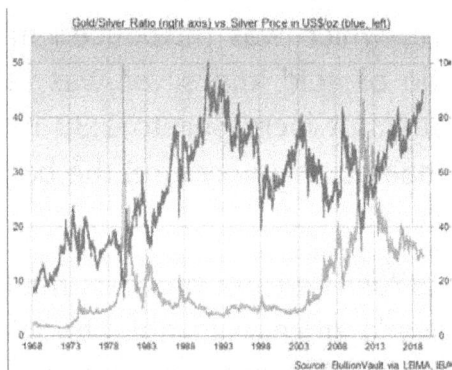

Gold/Silver Ratio (right axis) vs. Silver Price in US$/oz (blue, left)

Source: BullionVault via LBMA, IBA

When the ratio of gold to silver is greater than 40, you should begin buying larger quantities of silver than. If it falls below 40, it is recommended to begin to buy more silver and gold. This will help you come up with a plan to make the best from your investment in precious metals

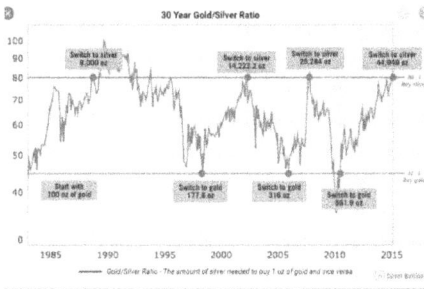

There are two pages more that always give me the gold-silver ratio at present.

This makes it clear why one should invest more in gold or silver.

13. How can I trade my silver and gold to earn a profit?

The time will come where you are able to sell your silver and gold and silver. You can, of course, attempt to sell your silver and gold in private. However, in the event of a crisis, it may be possible that you use your silver and gold products that are in demand and are able to be purchased using money that is very costly. It is beneficial to keep plenty of silver and gold.

A fair price is usually be negotiated for silver and scrap gold from refineries. The silver and gold is broken down, then given away, minus the cost for melting them down.

14. The way to go from part-time trading in silver and gold to becoming an all-time trader.

The other or one of them would prefer to make trading with precious metals their primary job. However, this should be thought through prior to making a decision. There are many regulations that vary according to the country. It is expected that these regulations will be extensive. My opinion is that private business has more prospects in this regard, as the demands for commercial sectors will rise dramatically. States aren't happy when their citizens put their money into precious metals. The value of these valuable metals is determined by market in the near future regardless of whether banks and states have intervened over and over time

Since I am able to purchase the equipment needed for testing on the internet, everyone is able to trade precious metals in a private way.

1. Register your business

The first recommendation for beginning your gold-based business is to register your business and get all the required licenses for business that are required for the trading of gold. You might want to consider registering your business in a limited liability corporation to shield the personal wealth of your clients from company liability.

2. Choose the type of gold you'd want to buy and then sell

Also, think about the kind of gold you would like to exchange. Do you wish to purchase old jewelry or damaged jewelry in order to refine them and sell them? Wedding rings, or brand new gold jewelry to resell or would you prefer trading with raw gold? Here are some areas of the gold trading industry that can be started now.

The purchase and sale of gold coins.

Jewelry purchases and sales auctions

Trade in used or broken jewelry Sales

The purchase and sale of old gold

3. Find out more about the latest trends in the field.

The gold industry goes beyond buying and selling gold . there are many new developments in the industry from which you can also profit. For instance, you can sell gold-plated products like mobile phones and tablet computers, or even luxurious wine bottles. You could even ask individuals to personalize your phones or even design your clothing with gold. These are all potential business opportunities to explore.

4. Learn more about gold's value

The gold business isn't something you can start as an aspiring. To become a successful gold dealer, you must be able to measure and weigh the quality of gold prior to purchasing it. It is essential to be aware of the signs of genuine gold, and also how to determine the value and cost of every piece at any given moment.

There is a common guideline to evaluate gold coins. However, gold jewelry should be evaluated based on its quality. There are tables standard to identify the quantity of gold contained in every piece, and also its worth in the marketplace.

Once you're aware the value of gold and feel that you could gain from it, purchase it. Be aware that prices for gold fluctuate significantly, which is why you have to be aware of the best time to purchase and sell. If you are buying wholesale,

You can purchase gold for less money and then sell it for the higher price.

5. Determine who the potential buyers of your gold could be.

Also, you must be able to select those who will purchase your gold. Check out the market to determine what the gold demand originates from, and who your rivals sell to.

It is also possible to form an association with jewelers around your area in order to offer them gold which they can then sell for the profit. They can search for gold wholesalers through which they can acquire many gold bars.

6. The development of an enterprise plan

Make sure you plan your business well and establish how you'll run your business as well as how you can grow your business in the coming years.

7. Purchase all the equipment you need.

It is also recommended to purchase all the equipment you need to begin your business, and perhaps some gold pieces to start. With a capital of $500 and a few hours of work, you could easily begin an initial business selling and buying gold.

8. Look for sources to purchase gold from

The most important aspect of this venture is the method and where you will get your gold. It is possible to create an office where those who want to sell their personal gold jewelry could come into and offer your gold to you. The gold you can purchase at a reduced price to allow you to sell for a profit in the future.

9. Promote and advertise your gold-based business

Like every other company, requires regular advertising and marketing to boost sales. If you have the money to pay for the cost of radio or television ads it's fine If you don't you can promote your business through the Internet or via classified ads in the local newspaper.

10. Secure and safeguard your company

The gold industry is a type of business that demands an extremely high level of security. A small amount of gold could be worth millions and if stolen or lost, it could cause a significant loss, therefore you have to consider the security aspects of your business carefully.

Apart from investing in a durable metal safe to store your items, you should hire security staff to guard your premises on regular basis.

11. Make a plan to grow and think big. your business

A different aspect to the business of gold you could look into is that customers come to swap your gold jewelry to another one for a small fee to be paid or people borrow cash from you taking your gold jewellery as collateral.

It is possible to earn a lot of profits by purchasing and selling gold. However, you should be sure to study everything you can about gold before you begin the business since it's an extremely risky business. Unaware novices can easily get caught in traps and fooled by scammers who sell gold.

It is advised that you collaborate with someone who has expertise in the field prior to purchasing as well as selling gold that you have purchased.

Chapter 11: The Urban Valley

Prospectors

This distribution is composed and intended to disclose the approaches to recovering gold from old PC and electronics.Therefore,It is encouraged to follow with normal sense.Follow generally wellbeing rules and laws.As data of topic is covered.Be mindful of the various security taking care of rules that differ from state to state.It is encouraged to counsel and investigate with their own consultant before any endeavor is begun in regards to individual situations.As this distribution isn't a preparation guide and doesn't clarify the treatment of dangerous poisonous materials and chemicals.

There are many different books and distributions as well as sites that can teach you the methods of

little by little the treatment of these dangerous materials.As as this isn't one of the books that deal with...

When deciding to proceed with this distribution, the reader agrees to release the distributer of any obligation which are suggested and revealed to pursue the goals depicted in this publication.

Method 1. Urban Valley Prospectors 3. Method 1. Equipment and Things.

There are only basic family synthetic chemicals that are associated with this type of training. These synthetics do not pose real dangers if they're not mixed. There are however certain risks that are present when we combine these substances, so taking certain security measures is essential to make sure you are using these synthetics. It is your responsibility to be sure to protect yourself whenever you use an amalgamation of synthetic substances. so by and large, make sure you are using a good judgement. In all likelihood, we must begin with our first step, and for this task you'll require a variety of products.

FIRST PRODUCT

Breathing equipment. There are a variety of cover materials available for purchase but for this purpose, you'll require one that is rated at least P100. Check out this particular rating system. Be sure to have crisp, clear air when you are in a closed space.

SECOND PRODUCT

Heavy duty gloves. Beware from the start the gloves made from latex or Nitrile. Make sure you use only PVC and rubber gloves.as these synthetics may adversely affect your skin and create problems when used for long periods of time.

THIRD PRODUCT

Safety Glasses. In the first place glasses should be safe and dissolvable. Anything else else isn't essential. Goggles or full face shields are equally effective. Make sure you wear something that is protective of the possibility of one mishap to cause visual impairment or even worse.

The REQUIRED Equipment

The exams can be conducted in a safer manner by using these items:

1. An old espresso maker works out fantastic or Borosil glass lab products like for example, a Pyrex bowl.

2. Two half gallon buckets.

3. Pick any of these containers, and then drill holes in the bottom. This can be used to create a strainer.

4. A 60 ounce pickle , or tea-type vessel with lid.

5. A box of filters for espresso.

6. A small oil pipe or the holder for the channel that can be removed from an espresso machine. 7. Spray bottle, some pipettes and a test tube.

8. Mail scale or food scale one that can measure Grams as well as Oz.

This is the Urban Valley Prospectors 4. Synthetics Required.

The most well-known synthetic compounds that are effective and can be discovered can be found in:

1. Gallon of Clorox Bleach. And then Sodium hypochlorate at 66 percent

2. Muriatic Acid HCL 34% Baume.

3. Hydrogen Peroxide - 3%, but not more noticeable than 5percent

4. Stump is removed by Bonide is available at Home Depot as well Sodium bisulfate 99.9 percent as per the label as Sodium Meta Bisulfite.

It is the Urban Valley Prospectors 5. The Selection of materials.

What are the primary materials from that you can extract gold

The basic amount of garbage from your products is the first step. There are a myriad of products that are gold-infused and various kinds of metals. Try to keep as significant portion of the waste out of the shower that is made of synthetic. This will make it easier to refine into that 24karet further down the line. The best and most simple materials are those found on expansion cards that are found in computer, and the "gold fingers" shown below are typically located on PCI cards Memory Sims/dims/ cards, etc. The CPU, also known by the name of Pentium Pro or Cyrix/AMD, or earlier 486 are an excellent option to recover the gold that was sucked out of. Particularly, these types of computers as well as older electronic models contain more gold that can be extracted through this method. Also, there is an electrical method that's suitable for removing precious gold off connects pins as well as similar objects as well as plated jewellery and similar items. This will be a separate publication coming soon. Keep an eye out.

6. This is the First Chemical Bath.

The base metals are made up of synthetic compounds, such as copper, zinc, and nickel. They are easily eliminated with the help of the compound bath.as the bath breaks down copper, letting the gold foils break off.

You must get rid of all visible non-gold-plated semiconductors, capacitors, and all remaining comparable parts in your computer, once you have finished with the plated parts. If you think you're seeing anything made of steel or iron take them out to improve the effectiveness of your results. This will result in lesser issues when you progress.

Tools are required

100 ML measuring glass or an espresso maker,

1 gallon bucket

Two 1/2 gallon buckets.

Chemicals are needed

1. Muratic is corrosive and

2. hydrogen peroxide.

Here are the resources you've been looking for.

The first step is to fill first the half gallon "with openings bored through sifter" pail filled with PC components you want to recover and put it in half gallon of wash water bucket.

In light of the fact that the interaction could result in dangerous gas, it is important to take your security equipment in a way that is unmatched on the basis that it only takes one incident to ruin your entire day with just a glance.

Second, as you monitor the level of muratic corrosive use, be sure to protect your fingers with gold a at least a good distance from the substances (1 centimeters is sufficient).

The next stage is to combine muratic corrosive and hydrogen peroxide in their proper quantities. Use a two-section (muratic corrosion) for 1-section (hydrogen peroxide) proportion to trigger the arrangement, and remove the gold foils. This will require approximately 5 percent of hydrogen peroxide.

This extraordinary mix is likely to be enough to breakdown the base metals.

that you've put in the first gallon.

It could be necessary to wait for anywhere between to 1 to two days for the entire cycle in order to finish.

For the best results, look at where the temperature ranges from between 80 and 90 degrees F.

At the end of the day, the mixture of synthetics should be a dull green hue and you'll need to look for a cluster of gold foils dangling from the surface within the gallon.

7. Filtering the Foils and Disposing waste.

Channels for espresso, the channel holder, and the 60oz pickle containers are required to remove the gold foils out of the acid that caused them to come about. Use the espresso channels to take every piece of gold before carefully pouring the corrosive out of the can of 1/2 gallon into the pickle container.

It is possible to do further cycles using the substance that remains So, put it in a separate bucket in a different bucket.

Now, you're engaged and halfway through the process and close to grabbing an gold button but you have to learn you must figure out how to handle the synthetic compounds that you have ran through the filter.

HOW DO I RECYCLE ACIDS

You can use a different compartment. I make use of void soft drinks in 2 liter containers or even old clothing containers to store the remaining acids. The acids can be placed in jugs with different amounts. Use a dark colored marker to label the containers where you've put the acids that are not needed. State"CLCu2"Toxic waste to remember. This will help with keeping track of the contents in that vast quantity of containers. Similar to the possibility of finding silver

Some of the synthetics are still in there which means you'll require a clutch or recuperation at a later time when you've advanced and a bit more skilled in gold recovery methods.

Everyone must be aware about the environment So, don't put the corrosive residues in the ground , or let it be a source of contamination for.

8. It is the Second Chemical Bath/Refining Foils.

Refine your gold foils

A veil of substance is required to complete this task.

The wastes generated by this process are extremely receptive consequently a valid security system is required.

Chemicals that are needed:

1. A 100 ml bottle.

2. Recovered foils.

3. Muratic acid.

4. Clorox.

The gold foils should be covered with a sufficient amount of Muratic corrosion. The gold thwarts will break down by adding the correct amount of Clorox using the pipette. To make the procedure more effective using glass rods.

There is an orange-colored arrangement towards the end of the process of dissolving.

To determine if there are any extra pieces of plastic that are green in the arrangement, it is necessary to channel it back using filtering coffee.

Filtering Particles

It is common to discover a handful of semiconductors, inductors and covers in the assembled arrangement given the fact that they are the ones that arrive in the final stage.

If you use an oversized flagon and neck pipe, you are able to make the arrangement again to create another arrangement that is a bright yellow hue.

9. How to Recover Your Gold.

It's probably the most thrilling part , and you'll enjoy a lot of fun reclaiming the gold in the solution

You'll require:

Plastic spoon.

-- Sodium Bisulfate and Stumpout.

The gold is likely to be dislodged synthetically in the structure because of the SO_2 gas which was produced in the course of to the addition of Sodium Bisulfate. The blend that was added after this mix is an opaque shade. It will also begin to release several bubbles.

The chlorine present in the solution is replaced with sulfur towards the end of this reaction. If you examine the lower region of the recepticle, you'll observe some earthy-colored hasten, which is gold that is dissolved within the reaction.

The final arrangement about 10 hours after when you've completed the exchange. The gold will be more comfortable during at this point. After that, it's completely settled. to run through the channel again, and that mud is your gold prior to the melting...

10. Method 2. Step-by-Step Gold Refining.

1. With a middle-punch using a mallet, you'll need to remove the covers from the clay-based CPU's starting point from the top before moving towards the bottom.

2. Inject the damaged CPU into a huge measuring utensil or an espresso maker.

3. The CPU's are covered with the basic amount of water from the tap.

4. After adding the water, bind it using an identical volume of HC1.

5. To prevent the fumes from escaping the pot to keep the fumes from leaving, cover it with a ceramic saucer.

6. It is not advisable to allow the water to reach an extreme point of boiling but it should remain under the temperature.

7. If you're using less than one pound of water, you need to add 1 teaspoon sodium nitrate and two teaspoons if the amount is greater.

8. It is recommended to remove the pot when the effervescing has stopped and the tiny air pockets have disappeared and the air pockets have disappeared. This should happen within about three quarters of an hour.

9. It is recommended to start the cycle starting from stage 3 if you observe salt in the mix and you are pouring dull corrosive into the channel once the mix has completely cool. There may be powder or gold thwarts on the channel. All solids must be placed within the vessel that responds to. The goal is to dissolve gold by condensing the stannous chloride by immersing it in acid.

10. In order to break down the metals in general it is recommended to rehash the stages 7 and 8 also.

11. It is really important to obtain only a quarter of the first volume or an unpleasant structure, which is why you have to get rid of the artistic saucer immediately. You are able to ignore the 11-13 stages if you use an appropriate amount of Nitric/Nitrate.

12. If you want to eliminate the amount of nitric, introduce HC1.

13. Redo the final two steps at least twice.

14. You can now make a fourth arrangement of ice 3D designs or weaken it using an equivalent amount in tap water.

15. To make sure that the structure free of suspended particles and silt, you must move cold through stuffed funnels under vacuum.

16. To obtain gold in an earthy powder structure, you'll should include Sodium Bisulfite to your solution.

17. The earthy colored powder captures siphon fluid inside its structure.

18. Mix everything thoroughly after you have added the required amount of water to over that brown powder.

19. Redo the two final stages at least twice.

20. The earthy colored powder should be covered with the crucial amount of the HC1(HC1 -

31 percent 10M muriatic acid.

21. It is possible to continue with the process the time that the shade of HC1 (in combination together with earthy-colored powder) stops obscuring the process of bubbling.

22. After you have removed the HC1 that is grimy You must try the next two steps till the excess HC1 stops staining. Use stannous chloride to clean the excess HC1 and then check whether there are any pieces of broken gold.

23. With a gentle shaking of the glass with the medium temperature it is important to

let the gold dry so that it is able to move freely.

24. Be careful not to soften the earthy-colored powder you'll want to transfer to a bowl to make it hot to red.

25. With the earthy-colored powder, you'll want to repeat every one of the strategies between 3 to 23.

26. Have a dissolving dish prepared.

27. In this dish for liquefying it is recommended to dissolve the gold powder that is dry.

11. Method 3. Step-by-Step Gold Refining.

The water regia technique calls for a container with the limit of 300ml per one ounce of scrap gold you want to refine. It is recommended to place the gold sullied in your plastic storage container...

The Aqua Regia method is to fill 3 pieces in the holders with corrosive hydrochloric then include one of nitric corrosion for the mix.

" OR, THIS WAY ASSURE"

Check the amount of ounces of metal in your possession and then add 30 milliliters of nitric corrosive per ounce compartment. Again, look at the quantity of ounces metal that are contained in the holder, and add 120 ml of hydrochloric muriatic corrosive per ounce.

If you want the process of disintegration process to be complete it is recommended to allow the gold disintegrate over a period of time, from an hour to overnight.

Utilizing a sifter made of pure material You should be able to sift particles of the corrosive structure that holds gold that is melting into an additional glass container.

Later on, after the gold has been broken. In the event you might find some important metals within the leftover particles, it is best to not to throw the rest away.

Based on the purity that the precious metal has, any excess corrosion that is present in the broken down gold must appear green in tone that ranges from golden and the color of emerald. Make use of a superior grade paper channel to channel the arrangement, assuming that you can see particles of any kind inside the container.

Mix the corrosive solution which contains the broken up gold with water or urea at a low speed. This will cause the broken down gold foam. The addition of water/urea is crucial due to how you do not require the corrosive to flow out of the container. The cycle should be stopped when you notice that the arrangement ceases to respond to the water or urea. Although this process will eliminate the nitric corrosive hydrochloric corrosive remains since it is pH has increased by 0.1 and up to 1.0.

It is evident that the arrangement turns brown as you start pouring sodium bisulfate into your container, and in a slow pace. It is important to understand that

the brownie mud that takes form in the container is pure gold, that's why that you must disrupt the order by using an unadulterated mixing rod made of glass.

In order to determine the gold answer it is essential to let the precipitant draw out all of the gold. You may need to add sodium bisulfate to the mix to ensure that the test comes positive.

Submerging the surface of the blending rod in the corrosive solution, you'll be tempted to test the results of the test to determine if it is gold. Create a wet stop pressing a wet towel against the edge of the pole that is used for blending. The wet spot should be covered with an ounce of Gold Recognition liquid.

If the spot turns into a brown/purple color, then it is a sign that gold is still dissolving within the acid. You should add more precipitant , or allow the amount that was previously added do its thing If you see these colors. At this point in this cycle entire amount of gold must be drawn out of the arrangement. The corrosion should have an obvious golden hue along with the earthy textured mud needs be in the bottom within the container.

You need to separate this earthy colored mud, and specifically the gold, by using the help of a fine-paper channel. Use this channel to pour the corrosive into another container or jug.

After you've completed pouring the corrosive into the mud, you need to add regular water to wash it thoroughly. Allow the mud to be able to settle before mixing it. Mix the corrosive from the compartment with the water from the faucet using the fine channel of paper. Keep the earthy-colored particles when you repeat this process at least a couple of times over a row.

Rinse the arrangements once more with salts that smell of water, then you can finish it off with regular water. You should add 10% alkali to 90% refined water. As you mix there will be white gaseous emissions. This process is crucial to get rid of the debasements in the gold mud and to kill the acids sticking on the gold mud.

Remove the mud from the fine paper channel by using an adequate water rinse.

Eliminate any excess fluid from your channel till it is nothing but clean golden mud. After that, melt the mud by placing the channel into a crucible.

There will be a metal appearance once the softening system is completed. You will receive 999.5 percent pure gold with no mishaps in the event that you have considered the methods and channels correctly. "PLATINUM"

The temperature at which water regia will not be enough to dissolve any platinum present inside your jewelry. When you pour out all the liquid, it is possible to observe that the platinum is left in the holder prior to the precipitation. Refining the platinum material again is essential to obtain it with high-quality. Make use of a fresh water regia shower to place the material. For this scenario you will need to warm up the corrosive and allow it to simmer. It is possible to let the corrosive to get warm maybe even for two hours. It is important to complete each ounce of broken down platinum with an ampmonium chloride, that will produce an red mud. The iridium that is remembered for this combination will result in an agile piece of harder than platinum and then it transforms into a blue-dark mud therefore you must persevere if you are planning to use to use it for the platinum. If you perform the stannous chloride test you might also find platinum group metals. There's a color assigned to each metal

Platinum turns red

Palladium transforms into

orange Iridium - transforms into

blue-black

12. Method 4. Step-by-step Refining Gold.

1. Use the circuit sheets of the piece and remove the gold bearing associations as well as edges

2. Make use of glass Pyrex compartment or an espresso maker to set the gold connectors

3. To get the necessary volume of water in an alternative glass Pyrex holder You

It is necessary to mix a mixture of hydrochloric corrosive and nitric corrosion in 1/3 ratio. Depending on the acids you centralize you will notice that the proportions will vary.

4. Place the gold connectors inside the corrosive mixture and let it remain there for

an hour.

5. Within 60 minutes use an espresso channel and disperse with any pollutants or unbroken up particles, while emptying the corrosive into a different glass Pyrex container.

6. To allow the substance to be completely consumed by the corrosive it is essential to include an element of precipitation at an inactive rate. Zinc is the best option. To determine whether there's any gold present in the mix make sure to use the identification fluid correctly following the directions. If required, add more zinc in

the mix until the gold has been completely in the blend.

7. The corrosive cup you employ to eliminate the excess corrosive should include one teaspoon of baking pop. Put this mixture aside for an hour. 8. By using small plastic tubing, gently tap the corrosive off the top without getting to the much-colored mud in the bottom of the holders. Mix this earthy colored mud with water that has been refined. Then, tap the water in the same holder that was removed corrosive after an hour.

9. The final step is the process of dissolving.

13. Method 5. Step-by-step Gold Refining.

There are many chemicals are used in the kitchen, in the laundry or in the garage each day However, when we understand to be cautious and use in a manner that is based on common sense, we can make our use safer and less complicated even when talking about hydrochloric or nitric acids. If used off-base, the corresponding chemicals can be extremely dangerous: channel unblockers solid dye pool synthetics salts and tree stump executioners sulfuric acid from vehicles, and much more. A sense of security and care in the use of these synthetic substances will help everyone else in avoiding any issues. Care and a sound judgement is essential when trying to refine gold or silver.

Be always mindful of the environment and keep everything organized in case of mishaps.

Here are the items you'll need

You will require the following: A plastic spoon, an ingredient made of plastic

stick.

An plastic"turkey Baster" pipette.

A tiny plastic funnel.

A shower bottle made of plastic.

They're all a couple of dollars if you are lucky that you buy these from a modest retailer close to where you live. You could use empty zest containers to substitute for the tiny glass chambers that are required to complete the process.

Bubble water should replace one of the containers for zest during the exchange, so an suitable glass bowl is needed. The best option for this scenario is Pyrex because it can aid in warming the ingredients in the container that you are using for flavor. The other important thing you'll require is disposable plastic gloves. Select a smaller item like the ones employed in handling food.

Coffee channels are also needed. Be careful not to spend an more than you can afford on these products.

These are also essential products:

A Jug containing 70-70% strength Nitric

corrosive. A container containing 32 percent

Hydrochloric acid.

If you look up the local home improvement websites there are many businesses that offer these products. You can purchase a bottle with 500 ml of each product.

Other metals in items that contain gold could be broken down with the aid of nitric corrosive. The final handling must be crafted with precision and pure gold.

When any remaining foreign metal substances have been eliminated, the gold is stripped and cleaned using the help of a mix that includes hydrochloric corrosion and Nitric corrosion. The statement that follows is crucial to be considered carefully.

You must stay out of any contact with the skin of these acids as they are a devastating trademark. Wear the gloves made of plastic whenever you use these acids. It is important to protect your eyes as well throughout the process. A secure and safe environment can be achieved through careful consideration, alertness and a clear mind. Utilizing these acids for extended periods of time has given me the results I required to prevent any inconvenience If you notice any chemical substances on your surface, use clean water to wash the synthetic compounds.

Be sure to finish the cycle outside, as there will be small quantities in nitrogen dioxide gases.

Safety and security should be the top priorities.

You'll be able to avoid any kind of problem when you take care and follow the security measures.

Urea (pee) as well as Sodium Meta Bisulphite are also essential when you want to refine gold even though they're not a great choice in the area of silver.

The grocery stores that have an area for nursery or all nurseries for plants, are the locations where you can observe the manure of plants, known as Urea. You can also purchase Urea for just a few dollars. Shop at the stores that sell home fermenting systems to purchase sodium meta Bisulphite. This particular synthetic is extremely efficient in extracting yeast from home-made libation holders. It's also not going to cost a lot (o few dollars also). Because the gold that is crude must be transformed into a sparkling metal, we require something that can be used to complete the refinement set tools. The end result is an electric light that is controlled by propane that is available in any home improvement store. A self-lighting product will be extremely beneficial. The item usually comes with buttons that are basically used to regulate the power of the flame. It is extremely efficient when it comes to the softening of gold and silver and is available at every Jack of all trades' set of equipment. There are a number of items that are gold-based on the internet or make use of old pieces of jewelry.

A similar mix of nitric acid and water is thought to break down the impurities in metal. By dissolving the components of the chamber and the quantity of metal you have it is possible to determine the most important amount of the nitric corrosion. In any case the nitric corrosive needs to be added towards the finish, then introduce the water using the aid of a pipette made of plastic.

Make use of bubbling water for speedier process. Put the glass tube in an evaporation shower. Make sure to keep up with the response with bubbling water at any time the flow and ebb water starts to cool.

The fluid begins to turn blue due to the reaction that is caused by copper contaminants.

The earthy-colored nitrogen dioxide gas can also cause an obvious response. Avoid the gas emitted through the arrangement. When the percolating ceases it is possible to unwind the device and let the reaction happen unattended.

You should pour in an nitric corrosive after the cycle has ended. This step is required to make sure that all pollutants are broken down. In the event that they are not broken down, the system will start to gurgle again. Add nitric corrosive to the glass until there is no further metallic contaminants in the glass.

It's a good opportunity to separate this gold from blue water which is why you need to make it less soluble by adding water. After the gold has been able to settle in the lower portion of the holder, you can pour the arrangement into a different compartment but without removing the gold.

Mix in some dishwashing liquid to loosen the gold on the bottom inside the container.

Since the gold may eliminate some of the pollutants that are in the water, it is recommended to include more water in this arrangement, and then wait for the gold to settle. It's a great time of emptying the container into a different compartment, but you must be cautious to ensure that you don't loss any precious gold. The yellow liquid will have some gold once you've gone over the pouring and weakening techniques.

It is important to be aware that you may find small amounts of silver that has been broken up in the blue liquid if that you've employed gold gems or a different combination of silver and gold. The process mentioned above could be used to recover silver too.

In order to recover 99.9 100% pure silver, would like to deal with the present gold in a way that will eliminate any potential contaminants.

Given the way the corrosive nitric is gone We really want to create an earthy-colored acceleration on the lower portion inside the glass bottle. This particular boost will be the solid structure that was derived by breaking up the gold.

Make a big part of one spoonful of Sodium Meta Bisulphite and add it to the yellow fluid to form the hasten. It will show the earthy color that will cover the yellow liquid after only few minutes. The yellow fluid will start to clear and later the earthy-colored encourage will settle on the bottom in the container. It will happen after about two up to 3 minutes. This will reveal the method that the gold that has been broken down is removed from the extra arrangement. To make sure that the total quantity of gold is increased it is essential to allow the process to take place over a few hours.

There are two angles when you look at the glass container:

1. The bottom in the glass vessel will have something that is reminiscent of earth.

colored dirt and

2. Old yellow liquid will be completely crystal clear.

Make use of a different holder to drain the distinctive fluid, trying to avoid disturbing the silt's earthy color. The discarded fluid should be diluted with water and then hunt the most secure method for organizing it.

Use the glass bowl that you created bubbles to remove the dregs and excess fluid. Consolidate the silt by adding sufficient amounts of water. Once the residue is settled on the bottom within the bowl begin pouring out the most water you can. Repeat this process several times to clean the residue thoroughly. It is important to have completely dried out the residue after which the fluid will usually disappear leaving this glass container in a heated area after you have eliminated lots of water.

We'll get an old, speculative process to this section. The fact that gold will not ever degrade, rust or stain is possibly the best quality of this precious metal.

The acids used in this cycle has to have the have the ability to alter the structure of this stable and non-reactive metal. The gold is dissolving! To break down gold, people who were before us have discovered the perfect method to do this: they've joined one piece of nitric corrosion with three pieces of hydrochloric acid.

This particular corrosive mixture was highly appreciated by ancient people, and they considered using a distinctive title: Royal Water. The Latin people have come up with the interpretation into"Water Regia". Add the corrosive mix (nitric as well as hydrochloric) to the glass tubes within which you've placed the gold that has been washed.

The dissolving system is going to begin rapidly and the fluid will turn yellow. If all previous impurities have been removed efficiently, the process will be excellent. In the event that it isn't, the hue of the fluid could appear green or blue.

A shower that bubbles can also be extremely beneficial for speeding the process.

The gold must be fully broken down, so allow the reaction to occur until the point of no return. Add more water regia into the liquid on the off chance that gold wasn't fully disintegrated by the time the response ends. Be cautious when dealing with the yellow fluid as it's gold in liquid form. It is best not to waste the gold by any means.

In the next step, you'll need to channel the yellow liquid without paying attention to its appearance. Make use of a second glass tube unfilled in which you'll transfer the yellow fluid to an clean filter.

The glass tube that is created should be flawlessly sparkling and washed to continue the process. This is essential because a dull, earthy-colored material will be formed from the yellow liquid and will settle on the bottom inside the bottle. It is possible that the gold will stick on the side of the vessel when it's not properly cleaned and cleaned. Imagine how long it would take to recover the gold that was lost. To make sure that all of the gold has made it through the channel to the bottle of glass, make use of the water shower jug to clean all the yellow liquid from the filter.

Utilize the same amount of water to lessen the strength of that yellow liquid. The yellow fluid could contain some hints of nitric corrosion and we need to get rid of them before beginning to promote the gold. It is crucial to kill the nitric corrosive as it can degrade any earthy color remnant that forms in the process of precipitation. We should use Urea to prevent the gold dregs out of being reabsorbed into the yellow pattern and help it in settling down at the bottom inside the glass vessel. Use the urea squeeze to pressing at a slow and steady speed. It is possible to be sure that the corrosive nitric has been eliminated if the urea is responsible for some

It rises up in the yellow fluid. The gurgling stops and then pull the gold from the arrangement as an earthy color leftovers. The next step is to include a small amount of SMB to make an earthy dark. The shadiness you choose to use will begin becoming darker and eventually will reach the lower end in the frame. The fluid will remain uncolored after it has been removed from the dim substance.

It is possible to identify crude gold as the earthy-colored dregs that have settled in the bottom in the container. Use a different holder to store the unique fluid, but do so cautiously. At this point you must reduce and eliminate any hint of corrosive in the dregs that are earthy-colored This is why to pour fresh clean water to it, and let it sit for a while.

This procedure should be repeated three times.

Then, you'll need to obtain only the earthy silt using the smallest amount of water you can. The silt with a earthy color will eventually dry out after the water evaporates.

The interaction was explained in detail with great detail so that all those who are interested in refining gold will get the opportunity to experience it.

Here's a less complicated method of achieving the same result 1. To dissolve other metal pollutions, simply place the gold piece in the acid nitric

2. Make sure to hold the gold while separating the blue arrangement accumulated during the dissolving process.

3. The gold will dissolve completely by adding 1 ounce of nitric corrosive as well as 3 ounces hydrochloric acid.

4. Get rid of any contaminants that may be present through filtering

5. Make use of urea to kill remaining nitric acid

6. Get the earthy gold dregs by using SMB

7. Get rid of the residue and then let it dry out so that you'll have a brownish powder.

8. The softening process should produce metallic gold out of that brown dust. This is enough for now.I believe you have learned and gained from what was explained.

In this publication.And be grateful for the opportunity to learn ways to transform yourself into an urban

Chapter 12: The Three 9v Battery

Colloidal Ssilver Maker

1.Take three or four snap connectors for batteries, cut the wires on the ends, and connect them in Series black to red - black to crimson, crimson to black as demonstrated under. Turn the end over, then fold inwards and wrap in electric tape. Solder and reduce the tube If you'd like.

2.You'll have one lead that is black (negative) in one place, and one red (high-quality) leading on the other quit.

3.Attach your black or red Crocodile clips to the appropriate wire. Bend twine to hang over the glass, as illustrated.

4.if you are using silver wire is preferred, the most simple to make use of and the smallest method of assembling it is to wrap the tape around the battery using wires that lead, jogging away from the snapped-on battery connectors (see the pinnacle. Once the tape is firmly bonded, it is hung in the vicinity of batteries that are not positioned correctly to the wires made of silver, as proved without delay.. (evian as well as other bottles of water make a powerful batch at room temperature, but its quality decreases rapidly.)

Methods for fast bottled water are described below.

Never let electrodes come into contact since batteries are quick (makes many). When storing the device, remove of crocodile clips that contain silver to ensure that no shorts do not occur. You can use a 1 9-volt battery

*separate batteries from the Crocodile battery clip meeting prior to carrying through airports.

Espresso maker with failproof design warm water approach.

This process results in a very small and very compact particle that can be stored in UV shielded bottles for 12-month or longer, and without the silver fallout of particles falling their own plus cost.

The Drip-espresso maker requires a hot Plate in order to keep hot water. This method is extremely useful as it gets the water up to the point of boiling immediately and in less than one minute, you'll be producing the best quality colloidal silver that you can get. After cooling, it's placed in a nice, pourable box for filling bottles. It is necessary to put in another one since the heating components used in the faucet or bottled water will be corroded by minerals, causing them to leak back into the distilled water since it is heated.

It is essential to heat the water to boost the conductivity of distillated water, despite the fact that it appears to be the silver is a pale yellow and can be produced at the temperature of room. This method can take many hours, but. It can take 6-8 hours, depending on the amount of water used and the voltage employed. One sparkling nine-volt battery contained in a four-ounce glass of water will change color in 2 to 3 hours.

The espresso pot style below suggests an electric power source of 30 volts and silver wires are added to the base tip.

At a radio store, you could take your cord that you bought and select the tip where you could insert the Diameter wire into the hole on the end. Should you not be able to do this, then you'll have cut off the tip of the supply cable for strength before removing the cord's ends and attach the clip crocodiles.

If you notice silver drawing attention on the purple crocodile clips electrode, you're reversed. The red (+) facet sinters silver molecules, while that facet of black (-) facet draws lots of them.

12 volts of electricity can absorb up to two hours or more to produce the an end product. 30 volts at about half this point.

It's time to create some of the most compact, tightest colloidal silvers ever made. It's a lot better than anything you could buy at the market.

To ensure the longest shelf life Make sure to store the colloidal silver you collect in U.V. covered brown or blue glass bottles. Blue bottles are excellent because you can discover the color of the colloid by conserving as much light. Brown bottles can't allow you to see shade as such.

Bottles with blue Uv protection can be purchased on the internet and at your local health food store usually sells the brown bottles. The silver is an average shelf life of 12 months or more without any separation or loss of shade.

Below is an the traditional double boiler technique. Be careful not to get yourself burned!

Chapter 13: Make Swallow Silver By Combining The Recent-Distillated Water (2 Pot Approach)

1.area pots in the oven and distilled water into one container (enough to fill the heavy glass jar until the top) and faucet water in another pot. Make sure to use low heat so that you don't break your glass jar.

2.Bring each pot of water to a point where they are boiling.

3.faucet water heater Warm to the lowest possible place. Do not heat your water in a microwave to warm it. It alters the molecular form of meals and water. Remove your microwave and don't consume microwaved food!

4.Slowly pour the distilled water slowly into the glass (peanut butter jars are usually painted well). Pour the water slowly and cautiously in order not to break glass.

5.set this glass of distilled water in a saucepan of hot tap water (as demonstrated in the previous paragraph).

Be cautious and gentle since all glass, water and pots could be extremely

Quick, room-temperature technique

Utilizes any great, drinking water, about as effective in treating common illnesses as the distilled water method, but should be utilized within a few days. The length of the particles is usually higher, and is likely to impact its effectiveness against specific organisms. Distilled water is generally the best.

Response time may be fast The above batch takes shorter than 2 minutes. Cold water may produce particles that are not homologous in the size. Let the water reach ambient temperature or higher but first warm it (not to microwave it) and pour it into the designated CS glass for making.

This batch will be done in less than five minutes. The color appears milky-white. You'll be able to recognize it's done when your batch is completely transparent (absolutely opaque). It's possible you'll not be in a position to see your hands (or other things) in the transparent glass.

suggestions to remember when making use of Canadian gold dollars made of maple for the production of colloidal silver

You'll have to hang your silver dollar from the pink or fine crocodile clip and then into the water, but be sure to ensure that the crocodile clip itself dry. It is not necessary to sinter copper molecules into the water.

Because there is no sintering at the negative edge The smooth, conductive metal can be employed. A paperclip made of metallic typically performs well. Be sure to remember your batch when you leave your home or, when you return, your currency could have lost significant weight, and your batch could be a bit over-decorated.

Electrodes are the most efficient way to sinter silver molecules from on the violet (+) side. Rotate your electrodes when you make use of two silver wires, so that they are evenly sinter.

Stop the process of making the beautiful metallic silver from scratch.

Chapter 14: 6 Matters To Keep In Mind Before Dying Your Diy Coloidal Silver Generator

BEFORE THE COLOIDAL USE CAN BE DIY SOLIDAR

1. Pureness begins by putting "w" in water. Impurities in tap water eliminate silver's advantages and cause safety concerns

The minerals in tap water, including the ones that make up "tough" drinking water could interact with silver colloidal, and create silver particles that form a clump, or mix and disintegrate from the Liquid. The worst part is that the area of the floor that is required to convert into bioactive silver has been drastically diminished, rendering the self-made silver virtually unusable*.

Even the distilled water you purchase at the grocery store may contain plastics and other contaminants leaching into the water prior to the time you buy it. Only when it is controlled in a precise manner, is it possible to understand how much silver gets into the water after an amount of time and day has been exceeded with a Generator.

2. Accurately assessing attention

There isn't a reliable method of measuring silver attention outside of a reputable and accredited laboratory setting. (TDS meters currently do not take measurements of silver!) This is the biggest issue. The health industry as well as it's United States EPA issued a maximum safety limit of intake of silver in steps from day to 350 micrograms for a typical 70kg adult. Without a chemical evaluation device that specifically measures silver awareness there is no method to estimate the amount of Silver is consumed.

"TDS" meters and the complete dissolving solid meters won't continue to measure the amount of silver specifically. As an alternative, they evaluate an electrical conductivity in the water that tells you the approximate quantity of charged ions in the water. They also make use of mathematical calculations to calculate the amount of attention the total of all dissolving solids. You can even include sodium chloride desk salt in the the water to gain a higher "ppm" study using their tds meter.

3. Controlling Concentration

If there is no way to determine the amount of silver that was actually used in products, it's impossible to determine how much of silver is being consumed.

Silver generators claim that you can alter the way the amount of silver put into water. This is true only when you have a unique control of the water. Outside of a precisely controlled laboratory or production facility which continuously evaluates its water best and processes for filtration it is difficult to achieve consistency with every batch.

The control of the perception of silver created is more complicated than just how much of it is currently passing through and the length of time it takes. Facet reactions are often in conjunction of the water impurities mentioned previously, and in certain instances, the inability to regulate water chemistry can cause the generator to break apart the water molecules on top that of the electrolytes made from silver. If this is turning your head it's a good reminder of why it's probably best to leave the work to professionals!

4. Topics on particle lengths:

The bioactive type of silver is the most definitely charged form. For silver particles that are colloidal to be effective* they must be transformed to a bioactive form and this could take position on the surface within the particles.

That means that the most tiny particle size can impact an best bio-pastime. Particle length analysis is also essential in analyzing many different final-product performance parameters. It's impossible to confirm the size of a particle without a transmission electron microscope.

It is crucial to understand that the size of particles not only immediately correlates to effectivenessHowever, it is also related the protection of your body! The body has to eliminate any silver it is not making use of. Given that it's able to best make use of what's on surface of the particle, it is required to eliminate all relaxation and waste. Think of it as having the outside of a huge onion, but having to dispose of the remainder of the layers in comparison to using smaller onions that are all. It's not until this waste saturates the framework that toxicity could be attained (see the sixth point below).

5. It's powered by electricity!

The passage of huge amounts of electricity through water could result in electrocution risks. Although it may seem obvious however, there are a number of safety risks when passing it in the current day and age through water. In the absence of appropriate control mechanisms and backup systems in the event that the controls fail, devastating consequences could result should someone come back into contact with the electrolytes that are energized, such as silver metal, or the solutions , when the generator is switched off.

6. Don't be blue!

In excess, excessive concentrations and size (leading to huge pieces) of silver could cause toxic, hazardous conditions. Self-made silver is the issue that led to"the "blue person" to make his skin blue due to the argyria. He began taking hundreds of times the who and epa daily day to protect himself from the dangers of homemade silver, with unconfirmed particle size and concentration every throughout the day, for months. Silver is more not always better. It is essential to have the bio-energetic form of silver to gain immune Assisthealth benefits.

As a general rule, you should be cautious in the event that you're thinking about using the silver produced by a colloidal generator, as described above. If you're looking to buy your silver supplement, search for products and brands that are made by professionals that you can think about.

It is the NSF trademark is a indicator that you've located a company that follows standards of best methods of manufacturing (GMP) in their factory. This certificate is proof of an enterprise which has made an investment of time and money to ensure that every collection they produce is safe*.

Chapter 15: When Is The Ideal Time To Purchase Silver And Gold From Private Individuals?

The question is simple to answer. The best time to ask is whenever people require cash. Most people must pay their bills by the end of each month. When there is a recession there are many who have to lose their jobs, but need to make payments. When the government pumps massive amounts of cash into the economy and citizens get social benefits, they do not have to sell their silverware.

However, there will come a point that you must pay your bills or pay off your obligations. When that happens, the best moment to purchase silver and gold that people have. In the deflationary period, it's best to purchase silver and gold. Because after every deflation, there is inflation. However, it is also possible we're experiencing the midst of stagflation. It is

179

a uncommon form of stagflation due to the fact that we are experiencing inflation at the same time , we are experiencing a decrease in financial power (downturn). When this happens the prices of silver and gold rise , and they are also rare. In addition, here I locate sellers as they need to pay off their obligations. The cost of silver and gold will increase steadily.

This time of year and the period following Christmas is also a suitable time to invest in silver and gold for private use. Since in the new year you are often required to pay for expensive bills and therefore, people have to sell their silver and gold.

The best time to buy is usually when more people need to sell than buyers, as the cost of selling falls. However, when there is an increase in silver and gold prices and increasing numbers of people invest in silver and gold it is the right time to sell silver and gold to protect the profits that have accrued.

Conclusion

This book is designed to the private trader who wishes to safeguard his own interests by trading and buying of silver and gold for the upcoming economic crisis. Cash bans and supply limitations in the commercial silver and gold trade force us to move towards the privately-owned market.

www.ingramcontent.com/pod-product-compliance
Lightning Source LLC
Chambersburg PA
CBHW071226210326
41597CB00016B/1960

9781774854372